JN276340

数学版　これを英語で言えますか？
Let's speak Mathematics!

保江邦夫　著

エドワード・ネルソン　監修

ブルーバックス

カバー装幀／芦澤泰偉事務所
カバー・オブジェ／植木ヒトミ
オブジェ撮影／南　弘幸
目次・章扉／WORKS（丸田早智子）

監修者まえがき

Let's speak Mathematics! に挑戦する日本の読者の皆さんに監修者として一言述べさせていただきます。

通常の文章の場合と同じように,数式を英語で声を出して読み上げるときには英国流とアメリカ流の違いがあります。例えば,英国の数学者は 0 をいつも nought と発音しますが,アメリカの数学者が 0 を nought と読むことがあるのはそれが下付の添字になっているときだけで,nought は sub zero の同義語として用いられます。

Let's speak Mathematics! ではアメリカ流の発音を採用していますが,それはアメリカ流のものが英国流よりもより正しいからではなく,単に私たちに馴染みやすいものだという理由によります。

また,ここでご紹介する数式の英語発音は数学における「存在定理」のようなものであり,決して「一意性定理」ではないという点にご留意ください。つまり,数式の発音として確かにアメリカの数学者の間に広く普及(存在)している読み方を示してはいますが,その読み方が一意的であると主張しているわけではありません。実際,ひとつの数式を読み上げる英語の発音はほとんどたいてい複数存在しているのです。

さらに,発音しない発音,つまり休止(ポーズ pause)の取り方も大事になってきます。数式の一部を括弧でくくる

場合，確かに "open parentheses ... close parentheses" という言い方をしますが，これもポーズを使い分けるとより簡略にすることができます。例えば，$x + (y \cdot z)$ は x plus **** y times z と発音し，$(x+y) \cdot z$ は x plus y **** times z と発音します（ここで **** の部分はポーズがあることを示しています）。

Let's speak Mathematics! で取り上げられている数式の例は，数学や物理学に出てくる数式の典型的なパターンを示すものであり，皆さんがこれらのパターンを学び取ったならば，$\aleph_0$（aleph nought）つまり可算無限個の数式を英語で読み上げることができるようになるのです。このように有限の大きさの投資が無限の大きさの収益を生むことができる努力の場は，他にはあまりないでしょう。

さあ，日本の皆さん，大いに Let's speak Mathematics! を楽しんでください！

2002 年初春

米国プリンストン大学数学科教授
エドワード・ネルソン

$$\begin{pmatrix} \text{Professor Edward Nelson} \\ \text{Department of Mathematics} \\ \text{Princeton University} \end{pmatrix}$$

はじめに

　絵画のように，方程式の美しさを理屈抜きで鑑賞してみようではないか！

　そんな雑文を月刊「数学セミナー」(日本評論社)に連載したのは，もうひと昔前のことです。これは，わりと好評だったようで，連載終了後に内容を膨らませた単行本『量子の道草——方程式のある風景——』(日本評論社)も出版されました。

　とはいっても，方程式を絵のように扱えば，数学などとは無縁の人たちにも方程式に興味を持ってもらえるのではないかと最初に気づいたのは僕ではありません（詳しいことは『量子の道草』まえがきとあとがき参照）。つまり，僕は人の褌で相撲をとったことになります。

　数式で描かれた方程式の場合，絵画とはいっても油絵を主体とする西欧絵画というよりは，中国や日本の水墨画に近い気がします。あるいは，本来それぞれが意味のある記号をつないで描き上げたものなのですから，芸術として鑑賞するなら「絵画」よりは「書」といった趣が強いともいえるでしょう。

　書という芸術は，むろん僕にはやはり絵画以上に遠い存在でしかありませんが，その理由は芸術としての書に書かれた漢字の羅列を判読して読み上げることさえ難しいからではないでしょうか。反面，たった一度だけではあっても

以前に耳にしたことのある漢詩ならば，それが勢いよく描かれた屏風は，書としての本来の輝きを増したものになるではありませんか。

その意味で，書は目だけで鑑賞する光の芸術ではなく，目と耳の両方で鑑賞する光と音の芸術といってよいでしょう。

ならば，方程式の美しさを鑑賞しつくすためには，目だけではなく耳にも訴える必要があるのではないでしょうか？

ということで，方程式を書の如くに読み上げながら愛でるというマニアックな芸術の高みを目指したのはいいのですが，どうもいまひとつピンときません。何かがおかしいというか，何かが足りないような気がして仕方がないのですが，それは書に描かれた漢詩を本来の中国語で読むのではなく，漢文として無理矢理日本語で読み上げる不自然さに近いものです。

やはり，漢詩は中国語で発音するのが本来の姿でしょうが，それならば方程式とて同じことがいえるのではないでしょうか。そもそも方程式に用いられる数式は欧米で作られ，我が国にはキリスト教と共に伝わってきたものでしかないのです。ということは，方程式を書の如く読み上げるとき，日本語で発音しても不自然でしかありません。

しからば，英語で方程式を読み上げながらその美しさを鑑賞すればよい！

そうはいったものの，たとえ慣れ親しんだはずの方程式も，いざ英語で発音しようと思ったとたんに遠い存在になってしまいます。

そう，英語でどう読めばいいのかが全くわからないのです。

はじめに

　責任転嫁するつもりはありませんが，そういえば高校から大学，大学院にいたるまで，ただの一度も数式の英語での読み方など教わったことはないのも事実です。

　外国の学会や研究会で外国人が数式を板書しながら英語で読み上げていたのを思い出すこともできますが，そこに出てくるのはかなり特化した専門分野でのみ多用される特殊なものが多いため，初等的な数式表現や方程式を英語で耳にしたことはなかったのです。

　といって，いまさらこの年で2次方程式の解の公式

　　にえーぶんのまいなすびーぷらすまいなするーと
　　びーじじょうまいなすよんえーしー

を英語ではどういうのかなど，辞書や英語の本を読んだり，英語をネイティブとする外国人に聞くなど，とてもできない相談です。

　しかし，自分として大いに興味ある事柄ではありますし，交流のある数学者や物理学者に聞いても，「確かに方程式の英語での読み方など誰も教えてはくれなかった，もし本が出るなら俺も買ってやる」という反応がほとんどでした。

　そして，「外国に長くいたんだし，国際的な学会で何度も研究発表やセミナーをうまくこなしていたんじゃないんだっけ。あの頃のことを思い出せばスラスラと出てくるんじゃないの？」というお世辞までも真に受けてしまったのです。

　その結果，数式の説明や方程式の解法など一行もないという，数学史上類い希な奇書，いや希書が誕生したのです。そこにあるのは，数式や方程式の英語での読み方の説明と，

昨 2001 年の国立大学入試に出た 100 を超える典型的な数式の英語発音の例示，それに大学に入ったら学ぶであろう物理や数学における著名な方程式の英語発音の例示のみ。

さあ，これまで数学など毛嫌いしてきた皆さん，これから外国留学を考えている皆さん，英語を学んでいる皆さん，大学受験生の皆さん，そして（これはごく少数でしょうが）将来数学者や物理学者になろうと考えている皆さん。どうか，世界に一冊しかないこの本で

>Let's speak Mathematics!

としゃれ込もうではありませんか。

もちろん，いくら外国で数学者や物理学者相手にセミナーや講演を数こなしたとはいえ，英語ネイティブではない日本人の数式英語発音など信用できないと思われるかもしれません。そこで，かねてより個人的にご指導くださった数学者である米国 Princeton 大学の Edward Nelson 教授に（最も重要な）英語発音の部分を監修していただきました。ここに感謝の意を表したいと思います。

Analytic Vector や Markov Field など，あるいは最近では数学基礎論における Axiomatic Nonstandard Set Theory など数多くの素晴らしい数学理論を生み出し，また世界で最も美しく短い数学の論文の著者としても知られている Nelson 教授に監修していただけたことは，この類い希な本の価値をさらに大きなものとするに違いありません。

それに，Nelson 教授からいただいた「監修者まえがき」の最後の行からわかるように，教授は数学的に奥の深いこと

をサラリと言ってのけてしまう才能にも長けていらっしゃいます。実際，我々人間の思考は，有限個のパターンを組み合わせることによって実に無限に多くの結果を生み出すことができるのですが，その数学的な証明自体は現代数学の根幹に位置する大問題のひとつなのです。

Nelson 教授が言外におっしゃりたかったのは，Let's speak Mathematics! の後には

Let's study Mathematics!

があるということに違いありません。

2002 年 4 月

保江邦夫

監修者まえがき……………5

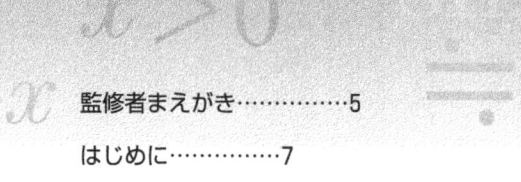

はじめに……………7

第I部　基礎訓練……………15

第1章　集合・写像・論理……………16

第2章　初等関数……………46

第3章　初等幾何……………63

第4章　数列と級数……………74

第5章　微分法……………83

第6章　積分法……………95

第7章　ベクトルと行列……………108

第8章　順列・組み合わせ・確率・統計……………121

第II部　実地訓練……………135

入試に出た数式の英語発音……………137～241

第III部 実践＝大学で学ぶ方程式 ……………243

相対性理論の方程式の英語発音……………245

古典力学の方程式の英語発音……………251

電磁気学の方程式の英語発音……………256

量子力学の方程式の英語発音……………262

場の量子論の方程式の英語発音……………268

素粒子論の方程式の英語発音……………273

流体力学の方程式の英語発音……………279

古典数学の方程式の英語発音……………284

英文索引……………290

和文索引……………296

第I部

基礎訓練

第1章 集合・写像・論理

まずは数学 mathematics の基本から Let's speak Mathematics! といきましょう。

要素[1] a が**集合**[2] A に属することを表す

$$a \in A$$

は

The element a is a member of the set A.

あるいは

a is an element of the set A.

や，より簡単に

a is a member of A.

とか

a is in A.

と読み，同じことを順序を逆に書いた式

$$A \ni a$$

は

---

[1] 要素 element
[2] 集合 set

The set A contains the element a.

あるいは簡単に

A contains a.

や

A includes a.

などと読みます。

何も要素のない集合は**空集合**[3] $\emptyset$ ですが，集合 B が空集合であることを意味する式

$$B = \emptyset$$

は

B equals the empty set.

と読みます。また，大文字の B であることを強調するときは

Upper case B is the empty set.

などといいます。

大文字と小文字については，省略することが多いのですが，あえて大文字であることを明示したいときは upper case あるいは capital をつけて発音します。例えば，大文字の E は

---

[3] 空集合 empty set

upper case *E*

あるいは

capital *E*

となります。小文字の場合には lower case や small をつけて

lower case *e*

あるいは

small *e*

となります。

数式では必ずといっていいほど**等号** =，即ちイコール記号が出てきます。例えば日本語では

$$左辺 = 右辺$$

のような等式の形となりますが，英語では[4]

$$\text{L.H.S.} = \text{R.H.S.}$$

で

The left-hand side equals the right-hand side.

と読みます。動詞の equal を使わずに形容詞の equal を使いたいときは

---

[4] 左辺 left-hand side (L.H.S.)，右辺 right-hand side (R.H.S.)

The left-hand side is equal to the right-hand side.

とします。ですから

$$A = B$$

は

A equals B.

でもよいし,

A is equal to B.

でもよいのです。

左辺と右辺が等しくないという式は

$$\text{L.H.S.} \neq \text{R.H.S.}$$

と書かれますが,これは

The left-hand side is not equal to the right-hand side.

あるいは

The left-hand side differs from the right-hand side.

などと発音します。例えば,ふたつの集合 A と B が等しくないことは

$$A \neq B$$

であり

    The set A differs from the set B.

あるいは

    A is not equal to B.

などと読まれます。

a が集合 B の要素ではないという式

$$a \notin B$$

は

    a is not a member of B.

と読みます。また，同じ式を

    B does not contain a.

あるいは

    a is not in B.

と読むこともできます。

次に部分集合 subset についてですが，集合 A が集合 X の部分集合であることを示す式

$$A \subset X$$

は

A is a subset of X

あるいは

A is included in X

や

A is contained in X

と読まれます。

同じ集合の包含関係を順序を変えて書いた式

$$X \supset A$$

は

X includes A.

あるいは

X contains A.

と読みます。

次に集合の間の演算についての式を発音しましょう。まずは集合 A と集合 B の**集合和** union が集合 C となる

$$A \cup B = C$$

は

The union of A and B equals C.

と読み，**集合積** intersection の場合

$$A \cap B = D$$

は

The intersection of A and B equals D.

となります。

何らかの**添字** suffix で区別された複数の集合についての全ての集合和

$$\bigcup_{k=1}^{n} A_k$$

は

The union from k equals one to n of A sub k.

といわれます。sub は subscript の省略形です。

1 は one ですが，しばしば unity とも呼ばれます。同様に

$$\bigcap_{j=0}^{\infty} B_j$$

は

The intersection from j equals zero to infinity of B sub j.

と読みます。0 は zero ですが null も使いますし，英国風に **nought** とやるのも目を引きますね。

また，集合和と集合積それぞれについての**交換法則**（交換律）commutative law

$$A \cup B = B \cup A$$
$$A \cap B = B \cap A$$

は

The union of A and B equals that of B and A, the intersection of A and B equals that of B and A.

と読めます。また，**結合法則**（結合律）associative law

$$(A \cup B) \cup C = A \cup (B \cup C)$$
$$(A \cap B) \cap C = A \cap (B \cap C)$$

は

The union of the union of A and B, and C equals the union of A, and the union of B and C, and the intersection of the intersection of A and B, and C equals the intersection of A, and the intersection of B and C.

となりますが，このように長い場合には集合和の記号 ∪ を cup，集合積の記号 ∩ を cap と呼ぶことにして

Parentheses A cup B close parentheses cup C equals A cup parentheses B cup C close parentheses, and parentheses A cap B close parentheses cap C equals A cap parentheses B cap C close parentheses.

と読むほうがポピュラーでしょう。

括弧（ ）は複数で parentheses ですが brackets も使います。単数形はそれぞれ parenthesis と bracket です。括弧には幾つかの種類があり，

$$\{\ \}$$

は

braces

で，単数形は brace です。また，

$$[\ ]$$

は

square brackets

ですが，誤解の心配がない場合には brackets だけでもかまいません。[ ] を square brackets というのに準じて（ ）を round brackets ということもあります。これらの括弧が混在しているような式として

$$[\{(A \cup B) \cap C\} \cup D] \cap E = \emptyset$$

を発音してみましょう。

Brackets braces parentheses A cup B close parentheses cap C close braces cup D close brackets cap E equals the empty set.

集合和と集合積の間の**分配法則**（分配律）distributive law

$$(A \cup B) \cap C = (A \cap C) \cup (B \cap C)$$
$$(A \cap B) \cup C = (A \cup C) \cap (B \cup C)$$

は

The intersection of the union of A and B, and C equals the union of the intersection of A and C, and the intersection of B and C. The union of the intersection of A and B, and C equals the intersection of the union of A and C, and the union of B and C.

と読みます。あるいは

A union B intersect C equals A intersect B union B intersect C. A intersect B union C equals A union C intersect B union C.

とすることもできます。

次に集合の**差** difference の演算を含む式

$$A \setminus B = C$$

ですが、これは

A setminus B equals C.

あるいは単に

A minus B equals C.

となります。また，**対称差** symmetric difference

$$A \triangle B$$

は

the symmetric difference of A and B

と読み，従って公式

$$A \triangle B = (A \cup B) \setminus (A \cap B)$$

は

The symmetric difference of A and B equals the union of A and B minus the intersection of A and B.

と発音します。

集合 X の部分集合 A の**補集合** complement

$$X \setminus A$$

は A^c と表されますが，発音は

A super c

で，super は superscript の省略形です。ですから

$$A^c \equiv X \setminus A$$

という補集合の定義式は

> The complement of the subset A of X is defined to be X minus A.

と読むのが正式ですが，簡単に

> A super c is defined to be X minus A.

でもかまいません。

いわゆるド・モルガンの公式 De Morgan's formula の一方

$$\left(\bigcup_\alpha A_\alpha\right)^c = \bigcap_\alpha A_\alpha^c$$

を発音すると

> The complement the union for all alpha of A sub alpha equals the intersection for all alpha of the complement of A sub alpha.

となり，もう一方

$$\left(\bigcap_\alpha A_\alpha\right)^c = \bigcup_\alpha A_\alpha^c$$

は

The complement the intersection for all alpha of A sub alpha equals the union for all alpha of the complement of A sub alpha.

となります。

集合の中でも，有限個の要素からなる**有限集合** finite set の場合は

$$A = \{a_1, a_2, a_3, a_4, a_5\}$$

のように全ての要素を書き出して中括弧でくくりますが，これは

A equals the set with elements a sub one, a sub two, a sub three, a sub four, a sub five.

と読みます。また，有限個とはいっても，要素の数が多い

$$\Omega = \{\alpha, \beta, \gamma, \cdots, \omega\}$$

のようなときには

Upper case omega equals the set with elements alpha, beta, gamma, and so on, omega.

となります。

これからもギリシャ文字が使われることがあるので，ここで文字の発音をまとめておきます。大文字と小文字の次にあるのが読み方と英語のつづりです。

A	α	アルファ	alpha
B	β	ベータ	beta
Γ	γ	ガンマ	gamma
Δ	δ	デルタ	delta
E	ϵ	イプシロン	epsilon
Z	ζ	ゼータ	zeta
H	η	イータ	eta
Θ	θ	シータ	theta
I	ι	イオタ	iota
K	κ	カッパ	kappa
Λ	λ	ラムダ	lambda
M	μ	ミュー	mu
N	ν	ニュー	nu
Ξ	ξ	クシィ	xi
O	o	オミクロン	omicron
Π	π	パイ	pi
P	ρ	ロー	rho
Σ	σ	シグマ	sigma
T	τ	タウ	tau
Υ	υ	ウプシロン	upsilon
Φ	ϕ	フィー	phi
X	χ	カイ	chi
Ψ	ψ	プサイ	psi
Ω	ω	オメガ	omega

有限集合でない場合には，要素を全て列挙して中括弧でくくるやり方は難しくなりますが，例えば**自然数** natural number の集合[5]

$$\bm{N} \equiv \{0, 1, 2, 3, \cdots\}$$

は

The set of natural numbers is defined to be the set with members zero, one, two, three, and so on.

となります。しかし，多くは

$$\bm{N} = \{n | \forall n \text{ natural number}\}$$

のように要素を代表する記号（いまの場合は n）を最初に書き，縦棒かコロンで区切って要素が満たすべき条件を書き下す方法が使われます。これは

Bold upper case N equals the set of all natural numbers n.

と読みます。

同様にして**整数** integer の集合

$$\bm{Z} = \{p | \forall p \text{ integer}\}$$

は

Bold upper case Z equals the set of all integers p.

---

[5] 通常，自然数の中に 0 は含めないが，数学基礎論等では最近 0 を含める傾向にある。

といい，**有理数** rational number の集合

$$\boldsymbol{Q} = \{r | \forall r \text{ rational number}\}$$

は

Bold upper case Q equals the set of all rational numbers r.

となります。

実数 real number の集合

$$\boldsymbol{R} = \{x | \forall x \text{ real number}\}$$

は

Bold upper case R equals the set of all real numbers x.

と読み，実数の集合と有理数の集合の差が**無理数** irrational number の集合であることを示す式

$$\boldsymbol{R} \setminus \boldsymbol{Q} = \{y | \forall y \text{ irrational number}\}$$

は

The set of real numbers minus the set of rational numbers equals the set of all irrational numbers y.

と読みます。

また，**複素数** complex number の集合

$$C \equiv \{z | \forall z \text{ complex number}\}$$
$$= \{z = x + iy | \forall x, y \in \boldsymbol{R}\}$$

は

Upper case bold C is defined to be the set of all complex numbers z, equal to the set of all z of the form x plus $i\,y$ for any real numbers x and y.

となります。

集合の間の演算についての最後は**直積** Cartesian product です。集合 A と集合 B の直積

$$A \times B$$

は

the Cartesian product of A and B

と読みますが、集合の直積であることが明らかな場合は単に

A cross B

ということもあります。ちなみに直積集合の定義式

$$A \times B \equiv \{(a,b) | \forall a \in A, \forall b \in B\}$$

は

The Cartesian product of A and B is defined to be the set of all pairs $a\,b$ for any a in A and any b in B.

と読みます。

また，同じ集合の直積である**べき集合** power set

$$A^2 = A \times A$$

は

A squared equals the Cartesian product of A and A.

また

$$A^3 = A \times A \times A$$

は

A cubed equals the Cartesian product of three A's.

と読みます。4次以上のべき集合は

$$B^n = \overbrace{B \times \cdots \times B}^{n}$$

のような式で表されることが多いのですが，これは

B to the n-th power equals the n-fold Cartesian product of B.

あるいは

B to the power of n equals the n-fold Cartesian product of B.

と読みます。べき集合であることが明らかなときは power は省略してもかまいません。つまり,

> B to the n equals the n-fold Cartesian product of B.

です。

1次元, 2次元, 3次元の**空間** space や, さらには一般に n 次元や無限次元の空間として考えられる実数の集合のべき集合

$$\boldsymbol{R}, \boldsymbol{R}^2, \boldsymbol{R}^3, \cdots, \boldsymbol{R}^n, \cdots, \boldsymbol{R}^\infty$$

は

> Bold capital R, bold capital R squared, bold capital R cubed, and so on, bold capital R to the n, and so on, bold capital R to the infinity.

と発音します。

多数の集合の直積を表すのに用いられる記号

$$\prod_{k=1}^{n} A_k$$

は

> the Cartesian product from k equals one to n of A sub k

と読みます。

集合についてはこのくらいにしておき，今度は**写像** mapping について Let's speak Mathematics! といきましょう。

f が集合 X から Y への写像であることを示す式

$$f : X \longrightarrow Y$$

は

 f is a mapping from X to Y.

あるいは

 f maps X to Y.

といいます。写像 mapping の代わりに**対応** correspondence や**関数** function を使うこともあります。

f が写像であることを示す別の式

$$f : X \ni x \longmapsto y \in Y$$

は

 f maps x in X to y in Y.

と読みますが，y が写像 f によって x が写される**像** image であることを強調する式

$$f : X \ni x \longmapsto f(x) \in Y$$

は

 f maps x in X to f of x in Y.

といいます。あるいは，対応関係の矢印を使わないで簡単に

$$y = f(x)$$

と記す場合は

　　y equals f of x.

と読みます。

　写像 g が集合 A から集合 B への上への写像，つまり**全射 surjection** であることを示す式

$$g : A \xrightarrow{onto} B$$

は

　　g is a mapping from A onto B.

あるいは

　　g maps A onto B.

と読みます。また，中への写像であることを示す式

$$g : A \xrightarrow{into} B$$

は

　　g is a mapping from A into B.

あるいは

　　g maps A into B.

となります。

さらに，写像 ϕ が 1 対 1 の場合，つまり**単射** injection となるときには

$$\phi : \Omega \xrightarrow{1:1} W$$

のような式になり

> Phi is a one to one mapping from upper case omega into W.

と発音します。1 対 1 かつ上への写像である**全単射** bijection の場合には

$$\phi : \Omega \xrightarrow{1:1, onto} W$$

という式で

> Phi is a one to one mapping from upper case omega onto W.

となります。

写像

$$f : X \ni x \longmapsto y \in Y$$

の逆，つまり**逆写像** inverse mapping

$$f^{-1} : Y \ni y \longmapsto x \in X$$

は

f inverse maps y in Y to x in X.

と読みます。ですから，逆写像が集合の演算を保存するという公式

$$f^{-1}(A \cup B) = f^{-1}(A) \cup f^{-1}(B)$$
$$f^{-1}(A \cap B) = f^{-1}(A) \cap f^{-1}(B)$$

は

f inverse of the union of A and B equals the union of f inverse of A and f inverse of B, f inverse of the intersection of A and B equals the intersection of f inverse of A and f inverse of B.

と発音します。

写像の最後は写像と写像の合成です。写像 $f : X \longrightarrow Y$ と写像 $g : Y \longrightarrow Z$ の合成写像 **composite mapping** を示す式

$$g \circ f : X \ni x \longmapsto g(f(x)) \in Z$$

は

The composite mapping of f and g maps x in X to g of f of x in Z.

あるいは

g composite with f maps x in X to g of f of x in Z.

と読み，さらに写像 $h: Z \longrightarrow W$ と合成した写像

$$h \circ g \circ f : X \ni x \longmapsto h(g(f(x))) \in W$$

は

> The composite mapping of f, g and h maps x in X to h of g of f of x in W.

あるいは

> The composition h g f maps x in X to h of g of f of x in W.

となります。

実数の集合 $\boldsymbol{R}$ からそれ自身への写像

$$f : \boldsymbol{R} \ni x \longmapsto y = f(x) \in \boldsymbol{R}$$

が**実関数**[6] $y = f(x)$ でしたが，この場合には

> The function f maps x in the set of real numbers to y equals f of x in the set of real numbers.

と読み，$y = f(x)$ は単に

> y equals f of x.

と読みます。

複素関数[7] $w = f(z)$ のときは

---

[6] 実関数 real function
[7] 複素関数 complex function

$$f : C \ni z \longmapsto w = f(z) \in C$$

ですが,これは

> The function f maps z in the set of complex numbers to w equals f of z in the set of complex numbers.

となります。

写像の次は論理式で Let's speak Mathematics! です。

まず,ふたつの**命題** proposition が互いに**同値** equivalent であることを表す式

$$A \equiv B$$

ですが,左辺も右辺も命題であることを強調して読むと

> The proposition A is equivalent to the proposition B.

となりますが,命題であることが明らかであれば単に

> A is equivalent to B.

としてかまいません。

ふたつの命題の間の**論理演算** logical operation を表す式を順次発音していくと,命題 A と B の**論理和** logical sum 「A または B」を示す

$$A \lor B$$

は

the logical sum of A and B

あるいは単に

A or B

となります。logical sum の代わりに disjunction を使うときもあります。**論理積** logical product「A かつ B」の式

$$A \wedge B$$

は

the logical product of A and B

あるいは単に

A and B

と読み，logical product を conjunction ということもあります。また，**含意** implication「A ならば B」を表す式

$$A \longrightarrow B$$

は

A implies B.

と発音します。

命題 A の**否定** negation「A でない」は

$$\neg A$$

と書かれますが,これは

the negation of A

あるいは

not A

と読みます。命題 A と B が同値 equivalent であることを示す論理式

$$A \equiv B$$

は

A is equivalent to B.

と発音しますが,A と B が同値であることの定義式

$$(A \longrightarrow B) \wedge (B \longrightarrow A)$$

は

A implies B and B implies A.

となります。

x を何らかの集合 X の要素として,そのような x についての何らかの命題 $P(x)$ について,「任意の x に対して命題 $P(x)$ が成り立つ」という全称命題

$$\forall x P(x)$$

は

For all x the proposition P of x holds.

あるいは

For all x P of x.

と発音します。また，「命題 $P(x)$ が成り立つような x が存在する」という存在命題

$$\exists x P(x)$$

は

There exists an x such that the propositon P of x holds.

あるいは

There exists x such that P of x.

となります。さらに，「命題 $P(x)$ が成り立つような x が唯ひとつ存在する」という一意存在命題

$$\exists_1 x P(x)$$

は

There exists a unique x such that the propositon P of x holds.

あるいは

There exists a unique x such that P of x.

と読みます。

初等的な論理学に出てくる二重否定が肯定となることを示す式

$$\neg\neg A \longrightarrow A$$

は

The negation of the negation of A implies A.

あるいは

Not not A implies A.

と発音します。また、やはり初等論理学で有名なド・モルガンの公式 De Morgan's formula

$$\neg(A \vee B) \equiv (\neg A) \wedge (\neg B)$$
$$\neg(A \wedge B) \equiv (\neg A) \vee (\neg B)$$

は

The negation of A or B is equivalent to the negation of A and the negation of B. The negation of A and

B is equivalent to the negation of A or the negation of B.

あるいは

Not A or B is equivalent to not A and not B. Not A and B is equivalent to not A or not B.

となります。

第2章 初等関数

今度は2次関数や三角関数に代表される**初等関数** elementary function で Let's speak Mathematics! といきましょう。

最もポピュラーなものは **2次関数** quadratic function についての **2次方程式** quadratic equation

$$ax^2 + bx + c = 0$$

の解の公式

$$x = \frac{-b \pm \sqrt{b^2 - 4ac}}{2a}$$

かもしれませんが、これは

x equals minus b plus or minus the square root of b squared minus four $a\,c$ over two a.

と読みます。ちなみに2次方程式自身は

$a\,x$ squared plus $b\,x$ plus c vanishes.

となりますが、vanishes という少し気取ったいい回しではなく

$a\,x$ squared plus $b\,x$ plus c equals zero.

が普通でしょう。

ここで2次方程式についてよく出てくる変形公式を発音

しておきます。

$$ax^2 + bx + c = a\left(x + \frac{b}{2a}\right)^2 - \frac{b^2 - 4ac}{4a}$$

a x squared plus b x plus c equals a times x plus b over two a squared minus b squared minus four a c over four a.

$$x^2 + (a+b)x + ab = (x+a)(x+b)$$

x squared plus a plus b times x plus a b equals x plus a times x plus b.

$$x^2 - c^2 = (x+c)(x-c)$$

x squared minus c squared equals x plus c times x minus c.

また，これらの公式を利用して得られる**恒等式** identity（変数のどんな値についても恒等的に成り立つ等式）

$$\sqrt{a + b + 2\sqrt{ab}} = \sqrt{a} + \sqrt{b}$$

は

The square root of a plus b plus two times the square root of a b equals the square root of a plus the square root of b.

と発音しますし，**相加平均** arithmetic mean（算術平均）と**相乗平均** geometric mean（幾何平均）に関する**不等式** inequality

$$\frac{a+b}{2} \geq \sqrt{ab}$$

は

a plus b over two is greater than or equal to the square root of a b.

と読みます。

3 次方程式 cubic equation や **4 次方程式** fourth order equation などは**因数分解** factorization で 1 次方程式や 2 次方程式に変形しますが，典型的な因数分解を発音しておきます。

$$a^3 \pm b^3 = (a \pm b)(a^2 \mp ab + b^2)$$

は

a cubed plus or minus b cubed equals a plus or minus b times a squared minus or plus a b plus b squared, respectively.

と読み，

$$x^4 + x^2 + 1 = (x^2 + x + 1)(x^2 - x + 1)$$

は

x to the fourth plus x squared plus one equals x squared plus x plus one times x squared minus x plus one.

です。

　集合や論理に比べて初等関数は比較的よく知られていますので，ここでは初等関数についての重要な公式を Let's speak Mathematics!

　まずは**多項式関数** polynomial function から。

$$p(x) = \sum_{k=0}^{n} a_k x^k$$

The polynomial function p of x equals the sum from k equals zero to n of a sub k times x to the k.

$$\left(\sum_{k=0}^{n} a_k x^k\right)\left(\sum_{j=0}^{m} b_j x^j\right) = \sum_{k=0}^{n}\sum_{j=0}^{m} a_k b_j x^{k+j}$$

The sum from k equals zero to n of a sub k times x to the k times the sum from j equals zero to m of b sub j times x to the j equals the sum from k equals zero to n of the sum from j equals zero to m of a sub k times b sub j times x to the k plus j.

次は三角関数 trigonometric function です。最初は三角関数の**加法定理** addition theorem と呼ばれる公式から。

$$\sin(A + B) = \sin A \cos B + \cos A \sin B$$

Sine A plus B equals sine A cosine B plus cosine A sine B.

$$\sin(A - B) = \sin A \cos B - \cos A \sin B$$

Sine A minus B equals sine A cosine B minus cosine A sine B.

$$\cos(A + B) = \cos A \cos B - \sin A \sin B$$

Cosine A plus B equals cosine A cosine B minus sine A sine B.

$$\cos(A - B) = \cos A \cos B + \sin A \sin B$$

Cosine A minus B equals cosine A cosine B plus sine A sine B.

$$\tan(A + B) = \frac{\tan A + \tan B}{1 - \tan A \tan B}$$

Tangent A plus B equals tangent A plus tangent B over one minus tangent A tangent B.

$$\tan(A - B) = \frac{\tan A - \tan B}{1 + \tan A \tan B}$$

Tangent A minus B equals tangent A minus tangent B over one plus tangent A tangent B.

$$\sin A + \sin B = 2 \sin\left(\frac{A+B}{2}\right) \cos\left(\frac{A-B}{2}\right)$$

Sine A plus sine B equals two sine of A plus B over two, times cosine of A minus B over two.

$$\sin A - \sin B = 2 \cos\left(\frac{A+B}{2}\right) \sin\left(\frac{A-B}{2}\right)$$

Sine A minus sine B equals two cosine of A plus B over two, times sine of A minus B over two.

$$\cos A + \cos B = 2 \cos\left(\frac{A+B}{2}\right) \cos\left(\frac{B-A}{2}\right)$$

Cosine A plus cosine B equals two cosine of A plus B over two, times cosine of B minus A over two.

$$\cos A - \cos B = 2 \sin\left(\frac{A+B}{2}\right) \sin\left(\frac{B-A}{2}\right)$$

Cosine A minus cosine B equals two sine of A plus B over two, times sine of B minus A over two.

おっと,三角関数といえば

$$\sin^2 \theta + \cos^2 \theta = 1$$

という重要な公式がありましたね。これは

Sine squared of theta plus cosine squared of theta equals one.

と読みますが,素朴に

Sine squared theta plus cosine squared theta equals one.

としても間違いではありません。また

The square of sine theta plus the square of cosine theta equals one.

と読むこともできます。

次は三角関数の倍角(つまり 2 倍角)の公式, 3 倍角の

公式，半角（つまり $\frac{1}{2}$ 倍角）の公式です。

$$\sin 2A = 2\sin A \cos A$$

Sine two A equals two sine A cosine A.

$$\cos 2A = \cos^2 A - \sin^2 A$$

Cosine two A equals cosine squared A minus sine squared A.

$$\sin 3A = 3\sin A - 4\sin^3 A$$

Sine three A equals three sine A minus four sine cubed A.

$$\cos 3A = 4\cos^3 A - 3\cos A$$

Cosine three A equals four cosine cubed A minus three cosine A.

$$\sin \frac{\alpha}{2} = \pm\sqrt{\frac{1}{2}(1 - \cos \alpha)}$$

Sine alpha over two equals plus or minus the square root of one half of one minus cosine alpha.

$$\cos\frac{\alpha}{2} = \pm\sqrt{\frac{1}{2}(1+\cos\alpha)}$$

Cosine alpha over two equals plus or minus the square root of one half of one plus cosine alpha.

今度は**指数関数** exponential function と**対数関数** logarithmic function ですが，準備として**指数法則** law of exponent を読んでおきましょう。

$$a^0 = 1$$

a to the zero equals one.

$$a^{-n} = \frac{1}{a^n}$$

a to the minus n equals one over a to the n.

$$\sqrt[n]{a^m} = \left(\sqrt[n]{a}\right)^m = a^{\frac{m}{n}}$$

The n-th root of a to the m equals parentheses the n-th root of a close parentheses to the m, equals a to the m over n.

$$a^n a^m = a^{n+m}$$

a to the n times a to the m equals a to the n plus m.

$$\frac{a^n}{a^m} = a^{n-m}$$

a to the n over a to the m equals a to the n minus m.

$$(a^n)^m = a^{nm}$$

a to the n to the m equals a to the n times m.

指数関数

$$y = e^x$$

は

y equals e to the power of x.

と発音しますが,もっと簡単に

y equals e to the x.

ともいえます。また,指数を表す exponential を用いて

$$y = \exp x$$

で指数関数を表すこともありますが,これは

y equals the exponential of x.

と読みます。

y equals the exponential x.

でもかまいません。

指数関数のこの書き方は，特に

$$y = \exp\left(\frac{\sqrt{(\beta^2 + \gamma^2)\,f(x)}}{|\delta|}\right)$$

のように指数関数の肩の上にくるべき数式が複雑な場合に役に立ちますが，これは

> y equals the exponential of the square root of parentheses beta squared plus gamma squared close parentheses times f of x over the absolute value of delta.

と発音します。

指数関数についての公式を読んでみると，次のようになります。

$$e^x e^y = e^{x+y}$$

e to the x times e to the y equals e to the x plus y.

$$(e^x)^a = e^{ax}$$

e to the x to the a equals e to the a times x.

特に**虚数単位**[1] $i = \sqrt{-1}$ によって指数関数と三角関数とをつなぐ**オイラーの公式** Euler's formula

$$e^{i\theta} = \cos\theta + i\sin\theta$$

は

e to the i theta equals cosine theta plus i sine theta.

となります。これを逆に解いた公式

$$\cos\theta = \frac{e^{i\theta} + e^{-i\theta}}{2}$$
$$\sin\theta = \frac{e^{i\theta} - e^{-i\theta}}{2i}$$

は

Cosine theta equals e to the i theta plus e to the minus i theta over two, sine theta equals e to the i theta minus e to the minus i theta over two i.

と読みます。

指数関数を組み合わせて得られる**双曲線関数** hyperbolic function についての定義式を読むと

$$\sinh x = \frac{e^x - e^{-x}}{2}$$

---

[1] 虚数単位 imaginary unit

The hyperbolic sine of x equals e to the x minus e to the minus x over two.

$$\cosh x = \frac{e^x + e^{-x}}{2}$$

The hyperbolic cosine of x equals e to the x plus e to the minus x over two.

$$\tanh x = \frac{\sinh x}{\cosh x}$$

The hyperbolic tangent of x equals the hyperbolic sine of x over the hyperbolic cosine of x.

$$\coth x = \frac{\cosh x}{\sinh x}$$

The hyperbolic cotangent of x equals the hyperbolic cosine of x over the hyperbolic sine of x.

$$\operatorname{sech} x = \frac{1}{\cosh x}$$

The hyperbolic secant of x equals one over the hyperbolic cosine of x.

$$\operatorname{csch} x = \frac{1}{\sinh x}$$

The hyperbolic cosecant of x equals one over the hyperbolic sine of x.

となります。

また,様々な公式を読み上げてみると次のようになります。

$$\cosh^2 \alpha - \sinh^2 \alpha = 1$$

The hyperbolic cosine squared of alpha minus the hyperbolic sine squared of alpha equals one.

$$\sinh(\alpha + \beta) = \sinh \alpha \cosh \beta + \cosh \alpha \sinh \beta$$

The hyperbolic sine of alpha plus beta equals the hyperbolic sine of alpha the hyperbolic cosine of beta plus the hyperbolic cosine of alpha the hyperbolic sine of beta.

$$\sinh(\alpha - \beta) = \sinh \alpha \cosh \beta - \cosh \alpha \sinh \beta$$

The hyperbolic sine of alpha minus beta equals the hyperbolic sine of alpha the hyperbolic cosine of beta minus the hyperbolic cosine of alpha the hyperbolic sine of beta.

$$\cosh(\alpha + \beta) = \cosh\alpha\cosh\beta + \sinh\alpha\sinh\beta$$

The hyperbolic cosine of alpha plus beta equals the hyperbolic cosine of alpha the hyperbolic cosine of beta plus the hyperbolic sine of alpha the hyperbolic sine of beta.

$$\cosh(\alpha - \beta) = \cosh\alpha\cosh\beta - \sinh\alpha\sinh\beta$$

The hyperbolic cosine of alpha minus beta equals the hyperbolic cosine of alpha the hyperbolic cosine of beta minus the hyperbolic sine of alpha the hyperbolic sine of beta.

$$\sinh 2x = 2\sinh x\cosh x$$

The hyperbolic sine of two x equals two times the hyperbolic sine of x times the hyperbolic cosine of x.

$$\cosh 2x = \cosh^2 x + \sinh^2 x$$

The hyperbolic cosine of two x equals the hyperbolic cosine squared of x plus the hyperbolic sine squared of x.

対数関数 logarithmic function では 10 を底とする**常用対数** common logarithm

$$\log_{10}$$

を

　　log base ten

と発音し，e を底とする**自然対数** natural logarithm

$$\log_e$$

は

　　log base e

と発音しますが，後者を簡単に

$$\log$$

と書き

　　log（ログ）

と読みます。対数関数

$$y = \log x, \quad (\forall x > 0)$$

は

y equals log x for all x positive.

と発音しますが，これについての幾つかの公式を読み上げておきましょう。

$$\log xy = \log x + \log y$$

Log x y equals log x plus log y.

$$\log \frac{x}{y} = \log x - \log y$$

Log x over y equals log x minus log y.

$$\log x^n = n \log x$$

Log x to the n equals n log x.

$$\log 1 = 0, \log e = 1$$

Log one equals zero, log e equals one.

第3章 初等幾何

さあ，次は**幾何学** geometry で Let's speak Mathematics! といきましょう。といってもほとんどが三角形についての公式や**座標** coordinate についての**解析幾何** analytic geometry の公式ですので，かなりポピュラーな印象があるのではないでしょうか。

まずは三角形からです。ひとつの**三角形** triangle にはみっつの**頂点** vertex（複数は vertices）がありますが，それを A, B, C とします。頂点 A に対面する辺[1] $\overline{BC}$ の**長さ** length を a，B に対面する辺 $\overline{CA}$ の長さを b，C に対面する辺 $\overline{AB}$ の長さを c とし，各頂点での**内角** interior angle の大きさをそれぞれ頂点と同じ記号 A, B, C とします。当然ながら三角形の内角の和は 180 度でした。式では

$$A + B + C = 180°$$

となり，

> A plus B plus C equals one hundred eighty degrees.

と読みます。

また，三角形の三辺の長さについての不等式条件

$$|b - c| < a < b + c$$

---

[1] 辺 edge

は

> a is between the absolute value of b minus c and b plus c, exclusive.

ですね。内角 A が**直角** right angle となるような**直角三角形** right triangle の場合にはピタゴラスの定理 Pythagorean theorem

$$a^2 = b^2 + c^2$$

が成り立ちますが,これは

> a squared equals b squared plus c squared.

と発音します。

三角形ですぐに頭に浮かぶのは、いわゆる**正弦定理** law of sines ですね。式では

$$\frac{a}{\sin A} = \frac{b}{\sin B} = \frac{c}{\sin C}$$

となり,発音は

> a over sine A equals b over sine B, equals c over sine C.

です。

その次に浮かんでくるのが**余弦定理** law of cosines

第 I 部 基礎訓練

$$a^2 = b^2 + c^2 - 2bc\cos A$$
$$b^2 = c^2 + a^2 - 2ca\cos B$$
$$c^2 = a^2 + b^2 - 2ab\cos C$$

ですが，これは

> a squared equals b squared plus c squared minus two $b\,c$ cosine A. b squared equals c squared plus a squared minus two $c\,a$ cosine B. c squared equals a squared plus b squared minus two $a\,b$ cosine C.

と読みます。

また，別の余弦定理

$$a = b\cos C + c\cos B$$
$$b = c\cos A + a\cos C$$
$$c = a\cos B + b\cos A$$

は

> a equals b cosine C plus c cosine B. b equals c cosine A plus a cosine C. c equals a cosine B plus b cosine A.

となります。

三角形の**面積** area は「底辺 base かける高さ height 割る 2」でしたから，式では面積を記号 S で表して

$$S = \frac{1}{2}ab\sin C = \frac{1}{2}bc\sin A = \frac{1}{2}ca\sin B$$

となり，発音すれば

> S equals one half a b sine C, equals one half b c sine A, equals one half c a sine B.

です。

面積についてはヘロンの公式 Heron's formula

$$S = \sqrt{s(s-a)(s-b)(s-c)}, \quad \left(s = \frac{a+b+c}{2}\right)$$

がありますが，これは

> S equals the square root of s times s minus a times s minus b times s minus c, where s is a plus b plus c over two.

と読みます。

三角形についてはこのくらいにして，次は座標を使った幾何学，つまりデカルト Descartes の解析幾何 Cartesian analytic geometry です。x 軸[2]と y 軸により**直交座標系** system of rectangular coordinates が与えられた座標平面上の 2 点[3]の座標が (x_1, y_1) と (x_2, y_2) のとき，2 点間の**距離**[4]

---

[2] x 軸 x axis
[3] 点 point
[4] 距離 distance

ℓ は

$$\ell = \sqrt{(x_2 - x_1)^2 + (y_2 - y_1)^2}$$

となりますが,これは

> ℓ equals the square root of x sub two minus x sub one squared plus y sub two minus y sub one squared.

と読みます。

直線 straight line を定める**方程式** equation

$$y = ax + b$$

は

> y equals a x plus b.

ですが,特に 2 点 $(x_1, y_1), (x_2, y_2)$ を通る直線の方程式

$$y = \frac{y_2 - y_1}{x_2 - x_1}(x - x_1) + y_1$$

は

> y equals y sub two minus y sub one over x sub two minus x sub one times x minus x sub one plus y sub one.

と読みます。

点 (x_1, y_1) から直線 $ax + by + c = 0$ に向かって引いた

垂線 perpendicular の長さ ℓ は

$$\ell = \frac{|ax_1 + by_1 + c|}{\sqrt{a^2 + b^2}}$$

で与えられますが，これは

> ℓ equals the absolute value of a x sub one plus b y sub one plus c over the square root of a squared plus b squared.

と発音します。

点 (x_1, y_1) を**中心** center とする**半径**[5] r の**円** circle を定める方程式

$$(x - x_1)^2 + (y - y_1)^2 = r^2$$

は

> x minus x sub one squared plus y minus y sub one squared equals r squared.

と読みます。また，直線や円以外の一般の**図形** figure は座標 (x, y) の何らかの**関数**[6] $f(x, y)$ によって

$$f(x, y) = 0$$

という形の方程式で与えられますが，これは

---

[5] 半径 radius（直径 diameter）
[6] 関数 function

第 I 部 基礎訓練

f of x y equals zero.

と発音します。

例えば，**楕円** ellipse を定める方程式は

$$\frac{x^2}{a^2} + \frac{y^2}{b^2} - 1 = 0$$

ですが，これを読むと

> x squared over a squared plus y squared over b squared minus one equals zero.

となり，**双曲線** hyperbola の方程式

$$\frac{x^2}{a^2} - \frac{y^2}{b^2} - 1 = 0$$

は

> x squared over a squared minus y squared over b squared minus one equals zero.

となります。また，方程式 $f(x,y)=0$ が 2 次方程式 $y=ax^2+b$ や $x=cy^2+d$ などの形になるものは**放物線** parabola を表しますが，これについての読みは明らかでしょう。

図形で囲まれた範囲は**領域** domain と呼ばれますが，領域は不等式で指定されました。直線 $y=ax+b$ を含めてそれよりも上側の領域は不等式

69

$$y \geq ax + b$$

で与えられますが，これは

y is greater than or equal to *a* *x* plus *b*.

と読みます。

一般に関数 $y = f(x)$ のグラフ graph が定める図形は集合

$$\{(x,y)|y = f(x), -\infty < x < \infty\}$$

と考えられますが，これは

The set of all pairs *x* *y*, where *y* equals *f* of *x* for *x* between minus infinity and infinity.

と発音します。このとき，グラフ $y = f(x)$ の下側の領域は不等式

$$y < f(x)$$

で定められ，

y is less than *f* of *x*.

と読みます。

立体図形 solid figure については x 軸と y 軸の両方に原点 origin で直交 orthogonal する z 軸を導入した**空間座標系** system of spacial coordinates を利用します。空間座標がそれぞれ (x_1, y_1, z_1) と (x_2, y_2, z_2) で与えられる 2 点間の距離 ℓ は

$$\ell = \sqrt{(x_2 - x_1)^2 + (y_2 - y_1)^2 + (z_2 - z_1)^2}$$

で

ℓ equals the square root of x sub two minus x sub one squared plus y sub two minus y sub one squared plus z sub two minus z sub one squared.

と読むのは既に見た**平面幾何** plane geometry の場合と同様です。

直線を定める方程式は

$$\frac{x - x_1}{a} = \frac{y - y_1}{b} = \frac{z - z_1}{c}$$

という**連立方程式** simultaneous equations となり，これは

x minus x sub one over a equals y minus y sub one over b, equals z minus z sub one over c.

と読みます。特に2点 $(x_1, y_1, z_1), (x_2, y_2, z_2)$ を通る直線の方程式

$$\frac{x - x_1}{x_2 - x_1} = \frac{y - y_1}{y_2 - y_1} = \frac{z - z_1}{z_2 - z_1}$$

は

x minus x sub one over x sub two minus x sub one equals y minus y sub one over y sub two minus y sub one, equals z minus z sub one over z sub two minus z sub one.

となります。

平面 plane を定める **1** 次方程式 linear equation

$$ax + by + cz + d = 0$$

は

a x plus b y plus c z plus d equals zero.

と読み，点 (x_1, y_1, z_1) からこの平面 $ax + by + cz + d = 0$ に向かって引いた垂線の長さ ℓ を求める公式

$$\ell = \frac{|ax_1 + by_1 + cz_1 + d|}{\sqrt{a^2 + b^2 + c^2}}$$

は

ℓ equals the absolute value of a x sub one plus b y sub one plus c z sub one plus d over the square root of a squared plus b squared plus c squared.

と発音します。

点 (x_1, y_1, z_1) を中心とする半径 r の球 sphere を定める方程式

$$(x - x_1)^2 + (y - y_1)^2 + (z - z_1)^2 = r^2$$

は

x minus x sub one squared plus y minus y sub one squared plus z minus z sub one squared equals r squared.

と読みます。

第4章 数列と級数

今度の Let's speak Mathematics! のテーマは**数列** sequence と**級数** series です。これらについては数式とその発音が複雑になってきますが，一定のパターンがありますのでどうか気楽に読んでいってください。

まずは級数ですが，式で書くと

$$\{a_n\}_{n=1}^{\infty}$$

のようになり，これは

the sequence a sub n from n equals one to infinity

と読みます。より簡単に

$$\{a_n\}$$

と書いて

the sequence a sub n

と読むこともあります。

あるいは，それぞれの**項** term を明記して

$$a_1, a_2, a_3, \cdots$$

と書く場合は

a sub one, a sub two, a sub three, and so on

と発音します。

等差級数 arithmetic series の**一般項** general term を与える式

$$a_n = a + (n-1)d$$

は

a sub n equals a plus n minus one times d.

と読みますが，**漸化式** recurrence formula では

$$a_{n+1} = a_n + d, \quad (n = 1, 2, 3, \cdots)$$

となり

a sub n plus one equals a sub n plus d, for n taking the values one, two, three, and so on.

と発音します。また，一般項までの**部分和** partial sum を与える式

$$S_n = \frac{n}{2}\{2a + (n-1)d\}$$

は

S sub n equals n over two times braces two a plus parentheses n minus one close parentheses times d close braces.

と読みますが，より簡単には

S sub n equals n over two times two a plus n minus one times d.

としてもかまいません。

等比級数 geometric series の一般項を与える式

$$a_n = ar^{n-1}$$

は

a sub n equals a times r to the n minus one.

と読み，漸化式

$$a_{n+1} = ra_n, \quad (n = 1, 2, 3, \cdots)$$

は

a sub n plus one equals r times a sub n, for n taking the values one, two, three, and so on.

と発音します。一般項までの部分和を与える式

$$S_n = \begin{cases} \frac{a(1-r^n)}{1-r} & (r \neq 1) \\ na & (r = 1) \end{cases}$$

は

S sub n equals a times one minus r to the n over one minus r for the values of r different from one, n times a for r equal to one.

となります。

等比級数の和の公式の応用としては、元金 a を利率 r の複利で n 年間預金した場合の元利合計の式

$$S_n = \frac{a(1+r)\{(1+r)^n - 1\}}{r}$$

がありますが、これは

S sub n equals a times one plus r times one plus r to the n minus one over r.

と発音します。

数列 $\{a_n\}$ についての和の記号の定義式

$$\sum_{k=1}^{n} a_k \equiv a_1 + a_2 + a_3 + \cdots + a_n$$

は

The sum from k equals one to n of a sub k is defined to be a sub one plus a sub two plus a sub three plus, and so on, plus a sub n.

と読みます。

ここで、簡単な和の公式を列挙しましょう。**階乗** factorial が出てきますが、その定義式

$$n! = 1 \times 2 \times 3 \times \cdots \times (n-1) \times n$$

を先に読んでおくと

n factorial equals one times two times three times, and so on, times n minus one times n.

となります。積の記号 $\prod$ を用いると

$$n! = \prod_{k=1}^{n} k$$

と書け

n factorial equals the product from k equals one to n of k.

と読みます。

$$\sum_{k=1}^{n} 1 = n$$

The sum from k equals one to n of one equals n.

$$\sum_{k=1}^{n} k = \frac{n(n+1)}{2}$$

The sum from k equals one to n of k equals n times n plus one over two.

$$\sum_{k=1}^{n} k^2 = \frac{n(n+1)(2n+1)}{6}$$

The sum from k equals one to n of k squared equals n times n plus one times two n plus one over six.

$$\sum_{k=1}^{n} k^3 = \frac{n^2(n+1)^2}{4}$$

The sum from k equals one to n of k cubed equals n squared times n plus one squared over four.

$$\sum_{k=1}^{n} k^4 = \frac{1}{30} n(n+1)(2n+1)(3n^2 + 3n - 1)$$

The sum from k equals one to n of k to the four equals one over thirty times n times n plus one times two n plus one times three n squared plus three n minus one.

$$\sum_{k=1}^{n} k(k+1) = \frac{n(n+1)(n+2)}{3}$$

The sum from k equals one to n of k times k plus one equals n times n plus one times n plus two over three.

$$\sum_{k=1}^{n} k(k+1)(k+2) = \frac{n(n+1)(n+2)(n+3)}{4}$$

The sum from k equals one to n of k times k plus one times k plus two equals n times n plus one times n plus two times n plus three over four.

$$\sum_{k=1}^{n} \frac{1}{k(k+1)} = \frac{n}{n+1}$$

The sum from k equals one to n of one over k times k plus one equals n over n plus one.

$$\sum_{k=1}^{n} \frac{1}{k(k+1)(k+2)} = \frac{n(n+3)}{4(n+1)(n+2)}$$

The sum from k equals one to n of one over k times k plus one times k plus two equals n times n plus three over four times n plus one times n plus two.

$$\sum_{k=1}^{n} \left(\frac{1}{k!} - \frac{1}{(k+1)!} \right) = 1 - \frac{1}{(n+1)!}$$

The sum from k equals one to n of one over k factorial minus one over k plus one factorial equals one minus one over n plus one factorial.

$$\sum_{k=1}^{n}\left\{\frac{1}{a_k}-\frac{1}{a_{k+1}}\right\}=\frac{1}{a_1}-\frac{1}{a_{n+1}}$$

The sum from k equals one to n of one over a sub k minus one over a sub k plus one equals one over a sub one minus one over a sub n plus one.

$$a_n = a_1 + \sum_{k=1}^{n-1}(a_{k+1}-a_k), \quad (n \geq 2)$$

a sub n equals a sub one plus the sum from k equals one to n minus one of a sub k plus one minus a sub k, for n greater than or equal to two.

数列と級数についての発音練習の最後に，名前だけは有名な**フィボナッチ数列**[1] $\{b_n\}$ の一般項を読んでおきましょう。

$$b_n = \frac{1}{\sqrt{5}}\left\{\left(\frac{1+\sqrt{5}}{2}\right)^n - \left(\frac{1-\sqrt{5}}{2}\right)^n\right\}$$

b sub n equals one over the square root of five times one plus the square root of five over two to the n minus one minus the square root of five over two

---

[1] フィボナッチ数列 the Fibonacci series

to the n.

第5章 微分法

さあ，いよいよ苦手な微分法 differentiation です。といっても，英語で読むのは何も難しいことはありません。まずは微分 differential の定義式からいきましょう。

$$f'(x) = \lim_{h \to 0} \frac{f(x+h) - f(x)}{h}$$

は

f prime of x equals the limit as h tends to zero of f of x plus h minus f of x over h.

と読みます。また，同じ極限 limit を別の書き方にしたもの

$$f'(x) = \lim_{z \to x} \frac{f(z) - f(x)}{z - x}$$

は

f prime of x equals the limit as z tends to x of f of z minus f of x over z minus x.

となります。

微分をライプニッツ流 Leibnizian style の記号で表すと

$$f'(x) = \frac{df(x)}{dx}$$

83

となりますが，これは

 f prime of *x* equals *d f* of *x d x*.

と読みます。簡単に表すと

$$f' = \frac{df}{dx}$$

で，

 f prime equals *d f d x*.

といいます。

2 階微分 second order differential

$$f''(x) = \frac{d^2 f(x)}{dx^2}$$

は

 f double prime of *x* equals *d* squared *f* of *x d x* squared.

と読み，**3 階微分** third order differential

$$f'''(x) = \frac{d^3 f(x)}{dx^3}$$

は

 f triple prime of *x* equals *d* cubed *f* of *x d x* cubed.

第 I 部 基礎訓練

と読みます。さらに、**n 階微分**[1]

$$f^{(n)}(x) = \frac{d^n f(x)}{dx^n}$$

は

  f n prime of x equals d to the n f of x d x to the n.

と発音します。

 微分についての一般的な公式を幾つか読んでおきましょう。

微分操作 differential operation が**線形演算** linear operation であることを示す公式

$$\frac{d}{dx}\{af(x) + bg(x)\} = a\frac{df(x)}{dx} + b\frac{dg(x)}{dx}$$

は

  d d x parentheses a times f of x plus b times g of x close parentheses equals a times d f of x d x plus b times d g of x d x.

と読みます。また、積の微分公式

$$\frac{d}{dx}\{f(x)g(x)\} = \frac{df(x)}{dx}g(x) + f(x)\frac{dg(x)}{dx}$$

は

---

[1] n 階微分 n-th order differential

$d\,d\,x\,f$ of x times g of x equals $d\,f$ of $x\,d\,x$ times g of x plus f of x times $d\,g$ of $x\,d\,x$.

となり，商の微分公式

$$\frac{d}{dx}\left\{\frac{f(x)}{g(x)}\right\} = \frac{\frac{df(x)}{dx}g(x) - \frac{dg(x)}{dx}f(x)}{g(x)^2}$$

は

$d\,d\,x\,f$ of x over g of x equals $d\,f$ of $x\,d\,x$ times g of x minus $d\,g$ of $x\,d\,x$ times f of x over g of x squared.

となります。

合成関数 composite function の微分公式

$$\frac{d}{dx}f(g(x)) = f'(g(x))\frac{dg(x)}{dx}$$

は

$d\,d\,x\,f$ of g of x equals f prime of g of x times $d\,g$ of $x\,d\,x$.

と発音します。

以下に初等関数についての微分公式の発音を列挙しておきましょう。

第Ⅰ部　基礎訓練

$$\frac{dx^n}{dx} = nx^{n-1}$$

$d\,x$ to the $n\,d\,x$ equals $n\,x$ to the n minus one.

$$\frac{de^x}{dx} = e^x$$

$d\,e$ to the $x\,d\,x$ equals e to the x.

$$\frac{da^x}{dx} = a^x \log a$$

$d\,a$ to the $x\,d\,x$ equals a to the x times $\log a$.

$$\frac{dx^x}{dx} = x^x(1 + \log x)$$

$d\,x$ to the $x\,d\,x$ equals x to the x times one plus $\log x$.

$$\frac{d\log x}{dx} = \frac{1}{x}$$

$d \log x\,d\,x$ equals one over x.

$$\frac{d}{dx}\sin x = \cos x$$

$d\,d\,x$ sine x equals cosine x.

$$\frac{d}{dx}\cos x = -\sin x$$

$d\,d\,x$ cosine x equals minus sine x.

$$\frac{d}{dx}\tan x = \frac{1}{\cos^2 x}$$

$d\,d\,x$ tangent x equals one over cosine squared x.

$$\frac{d}{dx}\sinh x = \cosh x$$

$d\,d\,x$ the hyperbolic sine of x equals the hyperbolic cosine of x.

$$\frac{d}{dx}\cosh x = \sinh x$$

$d\ d\ x$ the hyperbolic cosine of x equals the hyperbolic sine of x.

多変数 multivariable の関数の場合には，微分を**偏微分** partial differential で表します。高校の数学には出てきませんが，大学に入るとすぐに習います。それは，物理学や工学に登場する様々な**法則** law や**原理** principle の多くが偏微分を含む形で与えられているからです。これについては，その発音を第 III 部「実践＝大学で学ぶ方程式」で取り上げていますので，（まだ見ぬ）大学での勉強に思いを馳せてください。

ここでは，その準備として偏微分の読み方を示しておきましょう。**変数** variable の数が多くても面倒なので，**多変数関数** multivariable function としては 2 変数関数 $z = f(x, y)$ を想定します。3 変数関数や 4 変数関数，あるいは一般に n **変数関数**[2]になっても発音は同じことですので。

偏微分の定義式

$$\frac{\partial f(x,y)}{\partial x} = \lim_{\Delta x \to 0} \frac{f(x + \Delta x, y) - f(x, y)}{\Delta x}$$

は

The partial of f of x y with respect to x equals the limit as delta x tends to zero of f of x plus delta x y minus f of x y over delta x.

---
[2] n 変数関数 n-variable function

と読みます。「何々について」の「ついて」を表す with respect to を書き記すときには、簡単に w.r.t. や w r t で済ませてしまうこともありますが、あまりお勧めはしません。また、tends to の代わりに goes to を使うこともあります。

偏微分 partial differential の記号 ∂ はロシア語の d にあたる文字ですので、partial と発音する代わりに

$d\,f$ of $x\,y\,d\,x$ equals the limit as delta x tends to zero of f of x plus delta $x\,y$ minus f of $x\,y$ over delta x.

としてもかまいません。変数 y についての偏微分の定義式

$$\frac{\partial f(x,y)}{\partial y} = \lim_{\Delta y \to 0} \frac{f(x, y+\Delta y) - f(x,y)}{\Delta y}$$

も

The partial of f of $x\,y$ with respect to y equals the limit as delta y tends to zero of f of $x\,y$ plus delta y minus f of $x\,y$ over delta y.

あるいは

$d\,f$ of $x\,y\,d\,y$ equals the limit as delta y tends to zero of f of $x\,y$ plus delta y minus f of $x\,y$ over delta y.

と読みます。

高階 higher order の偏微分については、例えば

$$f_{xx}(x,y) = \frac{\partial^2 f(x,y)}{\partial x^2}$$

は

f sub *x* *x* of *x* *y* equals the second partial of *f* of *x* *y* with respect to *x*.

となり,

$$f_{xy}(x,y) = \frac{\partial^2 f(x,y)}{\partial x \partial y}$$

は

f sub *x* *y* of *x* *y* equals the second partial of *f* of *x* *y* with respect to *x* and *y*.

となります。2 階偏微分演算 second order partial differential operation

$$\Delta = \frac{\partial^2}{\partial x^2} + \frac{\partial^2}{\partial y^2}$$

はラプラシアン Laplacian と呼ばれますが,この式は

The Laplacian equals the second partial with respect to x plus the second partial with respect to y.

と読みます。

2 変数関数の微分を偏微分で書き表す公式

$$df(x,y) = \frac{\partial f(x,y)}{\partial x}dx + \frac{\partial f(x,y)}{\partial y}dy$$

は

> $d\, f$ of x y equals the partial of f of x y with respect to x $d\, x$ plus the partial of f of x y with respect to y $d\, y$.

となります。

微分についての最後として，この公式を一般の n 変数関数に拡張した公式を読んでおきましょう。それは

$$\begin{aligned}
&df(x_1, x_2, \cdots, x_n) \\
&= \frac{\partial f(x_1, x_2, \cdots, x_n)}{\partial x_1}dx_1 + \frac{\partial f(x_1, x_2, \cdots, x_n)}{\partial x_2}dx_2 \\
&\quad + \cdots + \frac{\partial f(x_1, x_2, \cdots, x_n)}{\partial x_n}dx_n
\end{aligned}$$

という複雑な式ですが，

> $d\, f$ of x sub one, x sub two, and so on, x sub n equals the partial of f of x sub one, x sub two, and so on, x sub n with respect to x sub one $d\, x$ sub one plus the partial of f of x sub one, x sub two, and so on, x sub n with respect to x sub two $d\, x$ sub two plus, and so on, plus the partial of f of x

sub one, x sub two, and so on, x sub n with respect to x sub n d x sub n.

と読みます。同じ公式を和の記号を使って書き表した式

$$df(x_1, x_2, \cdots, x_n) = \sum_{k=1}^{n} \frac{\partial f(x_1, x_2, \cdots, x_n)}{\partial x_k} dx_k$$

は

d f of x sub one, x sub two, and so on, x sub n equals the sum from k equals one to n of the partial of f of x sub one, x sub two, and so on, x sub n with respect to x sub k d x sub k.

となります。

このように,一般の多変数関数についての微分公式は多くの偏微分を含んだ複雑な形となり,それを書いたり発音するのも楽ではありません。そこで,多変数の場合の偏微分を簡単に書き示す記号(**ナブラ** nabla という)

$$\nabla = \left(\frac{\partial}{\partial x_1}, \frac{\partial}{\partial x_2}, \cdots, \frac{\partial}{\partial x_n} \right)$$
$$= (\nabla_1, \nabla_2, \cdots, \nabla_n)$$

を使うことが多くなります。これは

Nabla equals the partial with respect to x sub one, the partial with respect to x sub two, and so on, the partial with respect to x sub n, equals nabla sub one, nabla sub two, and so on, nabla sub n.

と発音します。最近ではナブラを単に del（デル）と読み，

Del equals the partial with respect to x sub one, the partial with respect to x sub two, and so on, the partial with respect to x sub n, equals del sub one, del sub two, and so on, del sub n.

とする場合が多いようです。

このナブラ記号を使えば，先ほどの多変数関数の微分公式は

$$df(x_1, x_2, \cdots, x_n) = \sum_{k=1}^{n} \nabla_k f(x_1, x_2, \cdots, x_n) dx_k$$

という簡単な式となり，発音も

d f of x sub one, x sub two, and so on, x sub n equals the sum from k equals one to n of del sub k f of x sub one, x sub two, and so on, x sub n times d x sub k.

となります。

第6章 積分法

微分法の次は**積分法** integration で Let's speak Mathematics! です。**積分** integral には**不定積分** indefinite integral と**定積分** definite integral とがありますが，まずは微分演算の**逆演算** inverse operation としての不定積分からです。

$$\frac{d}{dx}F(x) = f(x)$$

d d x capital *F* of *x* equals *f* of *x*.

という関係が成立するとき，関数 $y = F(x)$ のことを

$$F(x) = \int f(x)dx + C$$

Capital *F* of *x* equals the integral of *f* of *x* *d* *x* plus *C*.

と表し，関数 $y = f(x)$ の不定積分とか，**原始関数** primitive function と呼ばれます。C は何らかの**定数** constant なので，

Capital *F* of *x* equals the integral of *f* of *x* *d* *x* plus constant.

と読むこともできます。

以下に初等関数についての不定積分の公式と発音を列挙します。

$$\int x^n dx = \frac{x^{n+1}}{n+1} + C$$

The integral of x to the n d x equals x to the n plus one over n plus one plus constant.

$$\int f(x)^n \frac{df(x)}{dx} dx = \frac{f(x)^{n+1}}{n+1} + C$$

The integral of f of x to the n times d f of x d x d x equals f of x to the n plus one over n plus one plus constant.

$$\int \frac{1}{x} dx = \log|x| + C$$

The integral of one over x d x equals the log of the absolute value[1] of x plus constant.

$$\int \frac{f'(x)}{f(x)} dx = \log|f(x)| + C$$

---

[1] 絶対値 absolute value

The integral of f prime of x over f of x $d\,x$ equals the log of the absolute value of f of x plus constant.

$$\int \frac{1}{x^2 - a^2} dx = \frac{1}{2a} \log \left| \frac{x-a}{x+a} \right| + C, \quad (a > 0)$$

The integral of one over x squared minus a squared $d\,x$ equals one over two a times the log of the absolute value of x minus a over x plus a plus constant, for a positive.

$$\int \sin x\, dx = -\cos x + C$$

The integral of sine x $d\,x$ equals minus cosine x plus constant.

$$\int \cos x\, dx = \sin x + C$$

The integral of cosine x $d\,x$ equals sine x plus constant.

$$\int \sin^2 x\, dx = \frac{1}{2}\left(x - \frac{1}{2}\sin 2x\right) + C$$

The integral of sine squared $x\,d\,x$ equals one half of x minus one over two times sine two x plus constant.

$$\int \cos^2 x\,dx = \frac{1}{2}\left(x + \frac{1}{2}\sin 2x\right) + C$$

The integral of cosine squared $x\,d\,x$ equals one half of x plus one over two times sine two x plus constant.

$$\int \frac{1}{\cos^2 x}\,dx = \tan x + C$$

The integral of one over cosine squared $x\,d\,x$ equals tangent x plus constant.

$$\int \frac{1}{\sin^2 x}\,dx = \frac{1}{\tan x} + C$$

The integral of one over sine squared $x\,d\,x$ equals one over tangent x plus constant.

$$\int \tan x\,dx = -\log|\cos x| + C$$

The integral of tangent x d x equals minus the log of the absolute value of cosine x plus constant.

$$\int \frac{1}{\tan x} dx = \log|\sin x| + C$$

The integral of one over tangent x d x equals the log of the absolute value of sine x plus constant.

$$\int \frac{1}{\cos x} dx = \frac{1}{2} \log \left(\frac{1 + \sin x}{1 - \sin x} \right) + C$$

The integral of one over cosine x d x equals one half of the log of one plus sine x over one minus sine x plus constant.

$$\int \frac{1}{\sin x} dx = -\frac{1}{2} \log \left(\frac{1 + \cos x}{1 - \cos x} \right) + C$$

The integral of one over sine x d x equals minus one half of the log of one plus cosine x over one minus cosine x plus constant.

$$\int \sin^n x \cos x\, dx = \frac{\sin^{n+1} x}{n+1} + C$$

The integral of sine to the n x cosine x d x equals sine to the n plus one x over n plus one plus constant.

$$\int \cos^n x \sin x \, dx = -\frac{\cos^{n+1} x}{n+1} + C$$

The integral of cosine to the n x sine x d x equals minus cosine to the n plus one x over n plus one plus constant.

$$\int e^x dx = e^x + C$$

The integral of e to the x d x equals e to the x plus constant.

$$\int a^x dx = \frac{a^x}{\log a} + C, \quad (a > 0, a \neq 1)$$

The integral of a to the x d x equals a to the x over log a plus constant, for a positive and different from one.

$$\int \log x\, dx = x\log x - x + C$$

The integral of log x d x equals x times log x minus x plus constant.

不定積分の次は定積分です。**区間**[2] $[a, b]$ で**連続** continuous な関数 $y = f(x)$ の a から b までの定積分は

$$\int_a^b f(x) dx$$

と書かれ,

the integral from a to b of f of x d x

と発音します。

まず，定積分の定義式を読んでおきましょう。

$$\int_a^b f(x)dx \equiv \lim_{n\to\infty} \sum_{k=0}^{n-1} f\left(a + k\frac{b-a}{n}\right) \frac{b-a}{n}$$

The integral from a to b of f of x d x is defined to be the limit as n tends to infinity of the sum from k equals zero to n minus one of f of a plus k times b minus a over n times b minus a over n.

このようにして定められた積分は，次のような基本的な

---
[2] 区間 interval

性質を持っていました。

$$\int_a^b f(x)dx = -\int_b^a f(x)dx$$

The integral from a to b of f of x d x equals minus the integral from b to a of f of x d x.

$$\int_a^c f(x)dx = \int_a^b f(x)dx + \int_b^c f(x)dx$$

The integral from a to c of f of x d x equals the integral from a to b of f of x d x plus the integral from b to c of f of x d x.

また，定積分の**線形性** linearity を示す公式

$$\int_a^b \{kf(x) + mg(x)\}\,dx = k\int_a^b f(x)dx + m\int_a^b g(x)dx$$

は

The integral from a to b of k times f of x plus m times g of x d x equals k times the integral from a to b of f of x d x plus m times the integral from a to b of g of x d x.

と読みます。

関数の微分と積分が組み合わさった公式も幾つか読んで

おきましょう。

$$\int_a^b \frac{df(x)}{dx} dx = f(b) - f(a)$$

は

The integral from a to b of $d\,f$ of $x\,d\,x\,d\,x$ equals f of b minus f of a.

と発音します。また，解析学の基本定理 the fundamental theorem of calculus と称される公式

$$\frac{d}{dx}\int_a^x f(z)dz = f(x)$$

は

$d\,d\,x$ the integral from a to x of f of $z\,d\,z$ equals f of x.

となります。

これを複雑にした公式

$$\frac{d}{dx}\int_{h(x)}^{g(x)} f(z)dz = f(g(x))\frac{dg(x)}{dx} - f(h(x))\frac{dh(x)}{dx}$$

は

$d\,d\,x$ the integral from h of x to g of x of f of $z\,d\,z$ equals f of g of x times $d\,g$ of $x\,d\,x$ minus f of h of

x times $d\,h$ of $x\,d\,x$.

といいます。

微分と積分が組み合わさった公式の中で、最も有名なのは部分積分 integration by parts の公式

$$\int_a^b f(x)\frac{dg(x)}{dx}dx = f(b)g(b) - f(a)g(a) - \int_a^b \frac{df(x)}{dx}g(x)dx$$

ですね。これを読むと

The integral from a to b of f of x times $d\,g$ of $x\,d\,x$ $d\,x$ equals f of b times g of b minus f of a times g of a minus the integral from a to b of $d\,f$ of $x\,d\,x$ times g of $x\,d\,x$.

となります。

定積分については様々な不等式が知られていますが、

$$\left|\int_a^b f(x)dx\right| \leq \int_a^b |f(x)|dx, \quad (a < b)$$

は

The absolute value of the integral from a to b of f of $x\,d\,x$ is less than or equal to the integral from a to b of the absolute value of f of $x\,d\,x$, for a less than b.

と発音しますし、シュワルツの不等式 Schwarz' inequality

$$\left(\int_a^b f(x)g(x)dx\right)^2 \leq \int_a^b f(x)^2 dx \int_a^b g(x)^2 dx$$

は

The integral from a to b of f of x times g of x $d\,x$ squared is less than or equal to the integral from a to b of f of x squared $d\,x$ times the integral from a to b of g of x squared $d\,x$.

と読みます。

最後に，初等関数の定積分についての幾つかの公式を列挙しておきましょう．

$$\int_0^a \frac{1}{x^2+a^2}dx = \frac{\pi}{4a}$$

The integral from zero to a of one over x squared plus a squared $d\,x$ equals pi over four a.

$$\int_0^a \sqrt{a^2-x^2}dx = \frac{\pi}{4}a^2$$

The integral from zero to a of the square root of a squared minus x squared $d\,x$ equals pi over four times a squared.

$$\int_0^a \frac{1}{\sqrt{a^2 - x^2}} dx = \frac{\pi}{2}$$

The integral from zero to a of one over the square root of a squared minus x squared $d\,x$ equals pi over two.

$$\int_0^{2\pi} \sin kx \sin mx\, dx = \begin{cases} 0 & (k \neq m) \\ \pi & (k = m) \end{cases}$$

The integral from zero to two pi of sine $k\,x$ times sine $m\,x\,d\,x$ equals zero for k different from m, and pi for k equal to m.

$$\int_0^{2\pi} \cos kx \cos mx\, dx = \begin{cases} 0 & (k \neq m) \\ \pi & (k = m) \end{cases}$$

The integral from zero to two pi of cosine $k\,x$ times cosine $m\,x\,d\,x$ equals zero for k different from m, and pi for k equal to m.

$$\int_0^{\frac{\pi}{2}} \sin^{2m} x\, dx = \int_0^{\frac{\pi}{2}} \cos^{2m} x\, dx$$
$$= \frac{2m - 1}{2m} \frac{\pi}{2} \prod_{k=2}^{m} \frac{2k - 3}{2k - 2}$$

The integral from zero to pi over two of sine to the two m x d x equals the integral from zero to pi over two of cosine to the two m x d x, equals two m minus one over two m times pi over two times the product from k equals two to m of two k minus three over two k minus two.

$$\frac{\pi}{4} < \int_0^1 \sqrt{1-x^n}\,dx < 1, \quad (n > 2)$$

The integral from zero to one of the square root of one minus x to the n d x is greater than pi over four and less than one, for the values of n greater than two.

第7章　ベクトルと行列

ベクトル vector と行列 matrix（複数は matrices）は複数の成分に添え字がつくなど，表記が面倒になるのに応じて英語発音も難しくなってきます。それでも，一定のパターンや省略方法があるため，慣れてくればいたって簡単に Let's speak Mathematics! といけます。

まずは始点 P から終点 Q に向かう矢印としてベクトル $\vec{v}$ を定める式

$$\vec{v} = \overrightarrow{PQ}$$

を読むと

　　v equals the vector PQ.

となります。矢印の長さはベクトルの長さとか絶対値と呼ばれますが，記号

$$|\vec{v}|$$

や

$$|\overrightarrow{PQ}|$$

で表され

　　the norm of v

及び

the length of the vector PQ

と発音します。

ベクトルの和と差，それに**スカラー倍** scalar multiplication についての線形性を示す式

$$m(\vec{u} \pm \vec{v}) = m\vec{u} \pm m\vec{v}$$
$$(k \pm m)\vec{v} = k\vec{v} \pm m\vec{v}$$

は

m times u plus or minus v equals m times u plus or minus m times v. k plus or minus m times v equals k times v plus or minus m times v.

と読みます。また，任意のベクトル $\overrightarrow{PQ}$ が原点 O を始点とするふたつの**位置ベクトル** position vector の差で書き表せるという式

$$\overrightarrow{PQ} = \overrightarrow{OQ} - \overrightarrow{OP}$$

は

The vector PQ equals the vector OQ minus the vector OP.

となります。

ふたつのベクトルの間には**内積** inner product あるいは**スカラー積** scalar product と呼ばれる**かけ算** multiplication が定められましたが，それはふたつのベクトル $\vec{u}, \vec{v}$ のなす

角度 angle を θ ラジアン radian として

$$\vec{u} \cdot \vec{v} = |\vec{u}||\vec{v}| \cos\theta$$

となり，

> The inner product of u and v equals the norm of u times the norm of v times cosine theta.

と読みます。

空間に3次元の直交座標を想定してベクトルを**成分** component で表示すると

$$\vec{v} = (v_1, v_2, v_3)$$

などとなり，これは

> The vector v equals the triplet v sub one, v sub two, v sub three.

と読みます。あるいは，これを1行3列[1]の行列と見て

> The vector v equals the one by three matrix v sub one, v sub two, v sub three.

としてもかまいません。

さらに，3行1列の行列と見て

---

[1] 行 row 列 column

$$\vec{v} = \begin{pmatrix} v_1 \\ v_2 \\ v_3 \end{pmatrix}$$

と記し

The vector v equals the three by one matrix v sub one, v sub two, v sub three.

とすることもありますね。これはベクトルについての**線形変換** linear transformation を 3 行 3 列の行列で表す場合に便利ですが,それは

$$\begin{pmatrix} v'_1 \\ v'_2 \\ v'_3 \end{pmatrix} = \begin{pmatrix} A_{11} & A_{12} & A_{13} \\ A_{21} & A_{22} & A_{23} \\ A_{31} & A_{32} & A_{33} \end{pmatrix} \begin{pmatrix} v_1 \\ v_2 \\ v_3 \end{pmatrix}$$

となり,

The three by one matrix v sub one prime, v sub two prime, v sub three prime equals the three by three matrix A sub one one, A sub one two, A sub one three, A sub two one, A sub two two, A sub two three, A sub three one, A sub three two, A sub three three, times the three by one matrix v sub one, v sub two, v sub three.

と発音します。

ベクトルの内積を成分で書き表した式

$$\vec{u} \cdot \vec{v} = (u_1, u_2, u_3) \cdot (v_1, v_2, v_3)$$
$$= u_1 v_1 + u_2 v_2 + u_3 v_3$$
$$= \sum_{k=1}^{3} u_k v_k$$

は

The inner product of u and v equals the triplet u sub one, u sub two, u sub three dot the triplet v sub one, v sub two, v sub three, equals u sub one times v sub one plus u sub two times v sub two plus u sub three times v sub three, equals the sum from k equals one to three of u sub k times v sub k.

と読みます。これはまた，行列の積を使って

$$(u_1, u_2, u_3) \begin{pmatrix} v_1 \\ v_2 \\ v_3 \end{pmatrix} = u_1 v_1 + u_2 v_2 + u_3 v_3$$

となり，発音は

The one by three matrix u sub one, u sub two, u sub three times the three by one matrix v sub one, v sub two, v sub three equals u sub one times v sub one plus u sub two times v sub two plus u sub three times v sub three.

となります。これから内積を

$$\vec{u} \cdot \vec{v} = (u_1, u_2, u_3) \begin{pmatrix} v_1 \\ v_2 \\ v_3 \end{pmatrix}$$

と書くこともできます。これは

The inner product of u and v equals the one by three matrix u sub one, u sub two, u sub three times the three by one matrix v sub one, v sub two, v sub three.

と読みます。

また,ベクトルの長さを成分で表した式

$$|\vec{v}| = \sqrt{v_1^2 + v_2^2 + v_3^2}$$
$$= \sqrt{\sum_{k=1}^{3} v_k^2}$$

は

The norm of v equals the square root of v sub one squared plus v sub two squared plus v sub three squared, equals the square root of the sum from k equals one to three of v sub k squared.

と発音します。

以下にベクトルの内積や長さについての一般的な公式を列挙しておきます。

$$\vec{u} \cdot \vec{v} = \vec{v} \cdot \vec{u}$$

The inner product of u and v equals that of v and u.

$$\vec{u} \cdot (a\vec{v} + b\vec{w}) = a\vec{u} \cdot \vec{v} + b\vec{u} \cdot \vec{w}$$

The scalar product of u and parentheses a times v plus b times w close parentheses equals a times the scalar product of u and v plus b times the scalar product of u and w.

$$\vec{v} \cdot \vec{v} = |\vec{v}|^2$$

The inner product of v and v equals the norm of v squared.

また，これは大学に入ってから習うものですが，ふたつのベクトルの間には内積の他に**外積** outer product とか**ベクトル積** vector product と呼ばれるかけ算が定められます。ベクトルの内積はスカラーつまり数になりましたが，記号

$$\vec{u} \times \vec{v}$$

で表され

the vector product of u and v

と発音されるベクトル積はベクトルとなります。式で書くと

$$\vec{u} \times \vec{v} = (u_1, u_2, u_3) \times (v_1, v_2, v_3)$$
$$= (u_2 v_3 - u_3 v_2, u_3 v_1 - u_1 v_3, u_1 v_2 - u_2 v_1)$$

となり,

> The vector product of u and v equals the triplet u sub one, u sub two, u sub three times the triplet v sub one, v sub two, v sub three, equals the triplet u sub two times v sub three minus u sub three times v sub two, u sub three times v sub one minus u sub one times v sub three, u sub one times v sub two minus u sub two times v sub one.

と読みます。

ベクトル $\vec{u}$ と $\vec{v}$ のベクトル積 $\vec{u} \times \vec{v}$ で与えられるベクトルの長さはそれぞれのベクトルを辺とする**平行四辺形** parallelogram の面積に一致します。それは公式

$$|\vec{u} \times \vec{v}| = \sqrt{|\vec{u}|^2 |\vec{v}|^2 - (\vec{u} \cdot \vec{v})^2}$$

で求められますが,これは

> The norm of the vector product of u and v equals the square root of the norm of u squared times the norm of v squared minus the inner product of u and v squared.

と読みます。

次に行列ですが，既にベクトルそのものを行列と見たり，ベクトルの線形変換を与えるものとして**正方行列 square matrix** が登場してきました．

ここでは 2 行 2 列や 3 行 3 列の正方行列を例にとって，行列の和，差，積及び行列式についての一般的な公式を発音しておきます．より大きな次数の行列や，n 行 m 列の非正方行列についても，読み方については大きな違いはありません．

例えばふたつの正方行列 A と B の積 AB を与える式

$$\begin{pmatrix} A_{11} & A_{12} & A_{13} \\ A_{21} & A_{22} & A_{23} \\ A_{31} & A_{32} & A_{33} \end{pmatrix} \begin{pmatrix} B_{11} & B_{12} & B_{13} \\ B_{21} & B_{22} & B_{23} \\ B_{31} & B_{32} & B_{33} \end{pmatrix}$$
$$= \begin{pmatrix} A_{11}B_{11} + A_{12}B_{21} + A_{13}B_{31} & A_{11}B_{12} + A_{12}B_{22} + A_{13}B_{32} \\ A_{21}B_{11} + A_{22}B_{21} + A_{23}B_{31} & A_{21}B_{12} + A_{22}B_{22} + A_{23}B_{32} \\ A_{31}B_{11} + A_{32}B_{21} + A_{33}B_{31} & A_{31}B_{12} + A_{32}B_{22} + A_{33}B_{32} \end{pmatrix}$$
$$\begin{matrix} A_{11}B_{13} + A_{12}B_{23} + A_{13}B_{33} \\ A_{21}B_{13} + A_{22}B_{23} + A_{23}B_{33} \\ A_{31}B_{13} + A_{32}B_{23} + A_{33}B_{33} \end{matrix} \Big)$$

を読むと

The three by three matrix A sub one one, A sub one two, A sub one three, A sub two one, A sub two two, A sub two three, A sub three one, A sub three two, A sub three three times the three by three matrix B sub one one, B sub one two, B sub one three, B sub two one, B sub two two, B sub two three, B sub three one, B sub three two, B

sub three three equals the three by three matrix A sub one one B sub one one plus A sub one two B sub two one plus A sub one three B sub three one, A sub one one B sub one two plus A sub one two B sub two two plus A sub one three B sub three two, A sub one one B sub one three plus A sub one two B sub two three plus A sub one three B sub three three, A sub two one B sub one one plus A sub two two B sub two one plus A sub two three B sub three one, A sub two one B sub one two plus A sub two two B sub two two plus A sub two three B sub three two, A sub two one B sub one three plus A sub two two B sub two three plus A sub two three B sub three three, A sub three one B sub one one plus A sub three two B sub two one plus A sub three three B sub three one, A sub three one B sub one two plus A sub three two B sub two two plus A sub three three B sub three two, A sub three one B sub one three plus A sub three two B sub two three plus A sub three three B sub three three.

となります．これではあまりに長くなるので，積の行列 AB の成分を表す式

$$(AB)_{ij} = \sum_{k=1}^{3} A_{ik}B_{kj}, \quad (i,j = 1,2,3)$$

が便利です。これは

> The $i\ j$ component of the product of the matrices A and B equals the sum from k equals one to three of A sub $i\ k$ times B sub $k\ j$, for $i\ j$ taking values one, two, three.

と読みます。

また,正方行列の**行列式** determinant については,2行2列の場合の定義式

$$\begin{vmatrix} a & b \\ c & d \end{vmatrix} \equiv ad - bc$$

を発音すると

> The determinant of the two by two matrix a, b, c, d equals a times d minus b times c.

となります。

逆行列 inverse matrix を与える式

$$\begin{pmatrix} a & b \\ c & d \end{pmatrix}^{-1} = \frac{1}{\begin{vmatrix} a & b \\ c & d \end{vmatrix}} \begin{pmatrix} d & -b \\ -c & a \end{pmatrix}$$

は

The inverse of the two by two matrix a, b, c, d equals one over its determinat times the two by two matrix d, minus b, minus c, a.

と読み，z 軸を回転軸とする角度 θ ラジアンの**回転** rotation を表す線形変換の式

$$\begin{pmatrix} x' \\ y' \\ z' \end{pmatrix} = \begin{pmatrix} \cos\theta & -\sin\theta & 0 \\ \sin\theta & \cos\theta & 0 \\ 0 & 0 & 1 \end{pmatrix} \begin{pmatrix} x \\ y \\ z \end{pmatrix}$$

は

The three by one matrix x prime, y prime, z prime equals the three by three matrix cosine theta, minus sine theta, zero, sine theta, cosine theta, zero, zero, zero, one times the three by one matrix x, y, z.

となります。

ベクトルと行列についての Let's speak Mathematics! を有名なケーリー-ハミルトンの定理 Cayley-Hamilton's theorem の式でしめくくっておきましょう。

$$\begin{pmatrix} a & b \\ c & d \end{pmatrix}^2 - (a+d)\begin{pmatrix} a & b \\ c & d \end{pmatrix} + (ad-bc)\begin{pmatrix} 1 & 0 \\ 0 & 1 \end{pmatrix}$$
$$= \begin{pmatrix} 0 & 0 \\ 0 & 0 \end{pmatrix}$$

The two by two matrix a, b, c, d squared minus a

plus d times the two by two matrix a, b, c, d plus $a\,d$ minus $b\,c$ times the two by two unit matrix one, zero, zero, one equals the two by two zero matrix zero, zero, zero, zero.

第8章 順列・組み合わせ・確率・統計

さあ，いよいよ Let's speak Mathematics! の基礎訓練における最後のテーマ，**順列** permutation と**組み合わせ** combination, そして**確率** probability と**統計** statistics です。

相異なる n 個から異なる r 個を選んで1列に並べる場合の数が順列の数でしたが，それは記号

$$_nP_r$$

で表され

the number of permutations of n elements taken r at a time

と発音します。これは公式

$$_nP_r = n(n-1)\cdots(n-r+1) = \frac{n!}{(n-r)!}$$

によって求められますが，これを読むと

The number of permutations of n elements taken r at a time equals n times n minus one times, and so on, times n minus r plus one, equals n factorial over n minus r factorial.

となります。

特に n 個全てを並べる場合の数は

$$_nP_n = n(n-1)\cdots 2\cdot 1 = n!$$

のように n の階乗 $n!$ で与えられますが,これは

The number of permutations of n elements taken n at a time equals n times n minus one times, and so on, times two times one, equals n factorial.

と読みます。

次に,相異なる n 個から異なる r 個を選ぶ場合の数は組み合わせの数と呼ばれ,記号

$$_nC_r$$

で表され

the number of combinations of n elements taken r at a time

と読まれます。この場合の数を与える公式

$$_nC_r = \frac{_nP_r}{r!} = \frac{n!}{r!(n-r)!}$$

は

The number of combinations of n elements taken r at a time equals the number of permutations of n elements taken r at a time over r factorial, equals n factorial over r factorial times n minus r factorial.

と発音します。

　順列の数や組み合わせの数をいちいち the number of ……
と読むのは面倒なので，順列の数

$$_nP_r$$

を

　　the permutations of n taken r

と読んだり，

　　the permutations n r

と省略することもできますし，記号のアルファベットをそのまま発音して

　　$n\,P\,r$

と読むこともできます。

　組み合わせについても

$$_nC_r$$

を

　　the combinations of n taken r

とか

　　the combinations n r

あるいは

n choose r

としてもかまいません。

組み合わせについての公式を幾つか発音しておきます。

$$_nC_r = {_nC_{n-r}}$$

The combinations of n taken r equals the combinations of n taken n minus r.

$$_nC_r = {_{n-1}C_{r-1}} + {_{n-1}C_r}$$

The combinations of n taken r equals the combinations of n minus one taken r minus one plus the combinations of n minus one taken r.

$$_nC_p \ {_{n-p}C_q} = \frac{n!}{p!q!r!}, \quad (p+q+r=n)$$

The combinations of n taken p times the combinations of n minus p taken q equals n factorial over p factorial times q factorial times r factorial, for p plus q plus r equals n.

順列と組み合わせの最後に，**2 項定理** binomial theorem と**多項定理** polynomial theorem の式を読んでおきましょう。まず，2 項定理の式

第 I 部 基礎訓練

$$(a+b)^n = \sum_{r=0}^{n} {}_nC_r a^{n-r} b^r$$

は

a plus *b* to the *n* equals the sum from *r* equals zero to *n* of the number of combinations of *n* elements taken *r* at a time times *a* to the *n* minus *r* times *b* to the *r*.

と読みます。これから出てくる公式

$${}_nC_0 + {}_nC_1 + {}_nC_2 + \cdots + {}_nC_n = 2^n$$

は

The combinations of *n* taken zero plus the combinations of *n* taken one plus the combinations of *n* taken two plus, and so on, plus the combinations of *n* taken *n* equals two to the *n*.

あるいは

The *n* choose zero plus *n* choose one plus *n* choose two plus, and so on, plus *n* choose *n* equals two to the *n*.

と発音します。

また，多項定理の式

$$(a+b+c+\cdots)^n$$
$$= \sum_{p,q,r,\cdots \geq 0, p+q+r+\cdots=n} \frac{n!}{p!q!r!\cdots} a^p b^q c^r \cdots$$

は

a plus b plus c plus, and so on, to the n equals the sum for p, q, r, and so on, taking positive values such that p plus q plus r plus, and so on, equals n of n factorial over p factorial times q factorial times r factorial, and so on, times a to the p times b to the q times c to the r times, and so on.

と読みます。

順列と組み合わせの次は確率と統計です。確率では何らかの**事象** event をアルファベットの大文字で表します。事象 A が起こらないという事象は, 事象 A の**余事象** complementary event と呼ばれ

$$\overline{A}$$

あるいは

$$A^c$$

などと記されますが, これはどちらも

the complement of A

と読みます。

126

事象 A の確率は

$$P(A)$$

と表され

the probability of A

と発音します。確率が満たす条件

$$0 \leq P(A) \leq 1$$

は

The probability of the event A takes values between zero and one, inclusive.

と読みます。

余事象の確率を与える公式

$$P(A^c) = 1 - P(A)$$

は

The probability of the complement of A equals one minus the probability of A.

となります。

n 個の**排反事象** exclusive event についての**加法定理** addition theorem の式

$$P(A_1 \cup A_2 \cup A_3 \cup \cdots \cup A_n) = \sum_{k=1}^{n} P(A_k)$$

は

The probability of the union of A sub one, A sub two, A sub three, and so on, A sub n equals the sum from k equals one to n of the probability of A sub k.

と読みます。また,排反事象とはならない一般の事象についての加法定理の式

$$P(A \cup B) = P(A) + P(B) - P(A \cap B)$$

は

The probability of the union of A and B equals the probability of A plus the probability of B minus the probability of the intersection of A and B.

と読み

$$\begin{aligned}&P(A \cup B \cup C) \\ =\ & P(A) + P(B) + P(C) \\ & - P(A \cap B) - P(B \cap C) - P(C \cap A) + P(A \cap B \cap C)\end{aligned}$$

は

The probability of the union of A, B, C equals the probability of A plus the probability of B plus the probability of C minus the probability of the intersection of A and B minus the probability of the intersection of B and C minus the probability of the intersection of C and A plus the probability of A intersection B intersection C.

となります。

n 個の**独立事象** independent event についての**乗法定理** multiplication theorem の式

$$P(A_1 \cap A_2 \cap A_3 \cap \cdots \cap A_n) = \prod_{k=1}^{n} P(A_k)$$

は

The probability of the intersection of A sub one, A sub two, A sub three, and so on, A sub n equals the product from k equals one to n of the probability of A sub k.

と発音し,独立でない一般の事象についての乗法定理の式

$$P(A \cap B) = P(A)P(B|A)$$

は

The probability of the intersection of A and B equals the probability of A times the conditional probability of B given A.

と読み

$$P(A \cap B \cap C) = P(A)P(B|A)P(C|A \cap B)$$

は

The probability of the intersection of A, B, C equals the probability of A times the conditional probability of B given A times the conditional probability of C given the intersection of A and B.

となります。

確率 $p_1, p_2, \cdots\cdots, p_n$ で値 $x_1, x_2, \cdots\cdots, x_n$ が実現される**確率変数** random variable を X とすると，その**期待値** expectation を定める式

$$E(X) = \sum_{k=1}^{n} x_k p_k$$

は

The expectation of X equals the sum from k equals one to n of x sub k times p sub k.

と発音します。期待値と呼ばないで**平均値** mean value と呼ぶこともありますが，これは確率変数の実現値の全体につ

いての統計と関連してくるからです。

そこで，統計に関する数式を少し読んでおきましょう。

まず，**分散** variance を与える式

$$V(X) = \sum_{k=1}^{n} (x_k - E(X))^2 p_k$$

は

The variance of X equals the sum from k equals one to n of x sub k minus the mean value of X squared times p sub k.

と読み，**標準偏差** standard deviation の式

$$SD(X) = \sqrt{V(X)}$$

は

The standard deviation of X equals the square root of the variance of X.

となります。

また，平均値と分散の間の一般的な公式を列挙しておきましょう。

$$E(aX + b) = aE(X) + b$$

The expectation of a times X plus b equals a times the expectation of X plus b.

$$V(aX + b) = a^2 V(X)$$

The variance of a times X plus b equals a squared times the variance of X.

$$V(X) = E(X^2) - E(X)^2$$

The variance of X equals the mean value of X squared minus the mean value of X, squared.

最後に,連続な実数値を実現値とする確率変数 X が a 以上 b 以下の値を実現する確率を積分として定める式

$$P(a \leq X \leq b) = \int_a^b p(x)dx$$

を読むと

The probability of X taking values between a and b, inclusive equals the integral from a to b of p of x $d\,x$.

となります。ここで関数 $y = p(x)$ は**確率密度関数** probability density function と呼ばれ,例えば平均値が m で標準偏差が σ の**正規確率密度関数** normal probability density function

$$p(x) = \frac{1}{\sqrt{2\pi}\sigma} e^{-\frac{(x-m)^2}{2\sigma^2}}$$

は

p of x equals one over the square root of two pi times sigma times e to the minus x minus m squared over two sigma squared.

と発音します。

第II部
実地訓練

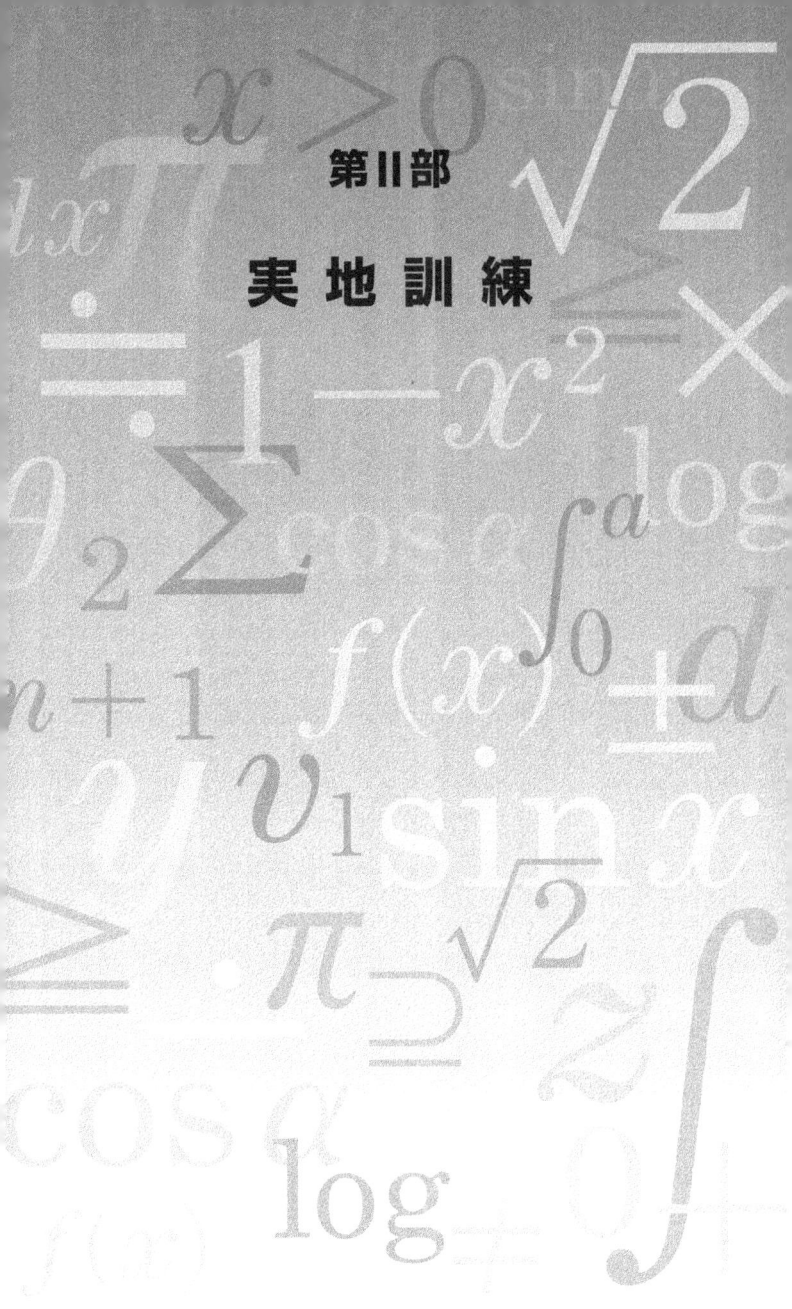

大学受験生の皆さんへ

これまでは Let's speak Mathematics! の基礎訓練として，高校で習う数学や，これから大学に入ってすぐに習う数学に登場してくる様々な数学公式や方程式の英語発音を学んできました。

今度は，Let's speak Mathematics! の実地訓練ということで，2001年に行われた国立大学入試で出題された数式を英語で読んでみましょう。もちろん，英語発音の練習としてだけでなく，数学解答の練習問題として見ておくこともいいかもしれません。

受験に必要な数学公式を憶えるのが苦手だと嘆く前に，ちょっと気分を変えて英語で数学公式を憶えてみるのもいいんじゃない？

入試に出た数式の英語発音 1

北海道大学

$$\int_0^a \frac{x^{2n+2}}{1-x^2} dx$$
$$= \frac{1}{2} \log \frac{1+a}{1-a} - \sum_{k=0}^{n} \frac{a^{2k+1}}{2k+1}$$

The integral from zero to a of x to the two n plus two over one minus x squared $d\,x$ equals one half the log of one plus a over one minus a minus the sum from k equals zero to n of a to the two k plus one over two k plus one.

入試に出た数式の英語発音 2

北海道大学

$$X = p \begin{pmatrix} 1 & 2 \\ 2 & 4 \end{pmatrix} + q \begin{pmatrix} 4 & -2 \\ -2 & 1 \end{pmatrix}$$

X equals p times the two by two matrix one, two, two, four plus q times the two by two matrix four, minus two, minus two, one.

入試に出た数式の英語発音 3
帯広畜産大学

$$B = \begin{pmatrix} 0 & 0 & 0 \\ a & 0 & 0 \\ \log_2 p & a - 2b & 0 \end{pmatrix}$$

B equals the three by three matrix zero, zero, zero, a, zero, zero, log base two of p, a minus two b, zero.

入試に出た数式の英語発音 4
旭川医科大学

$$t\sin\alpha = (1-t)\cos\alpha \quad \left(0 < \alpha < \frac{\pi}{2}\right)$$

t sine alpha equals one minus t cosine alpha, for alpha lying between zero and one half pi.

入試に出た数式の英語発音　5

室蘭工業大学

$$f(x) = 2x^3 + \int_0^2 f(t)dt,$$
$$\int_0^x (x-t)g(t)dt = \frac{1}{5}x^5 + \frac{1}{4}x^4 + kx^2$$

f of x equals two x cubed plus the integral from zero to two of f of t d t, the integral from zero to x of x minus t times g of t d t equals one fifth x to the fifth plus one quarter x to the fourth plus k x squared.

入試に出た数式の英語発音　6
室蘭工業大学

$$xy + \frac{4y}{x+1} + \frac{x}{y} + \frac{4}{(x+1)y} \geq 6$$

x y plus four y over x plus one plus x over y plus four over x plus one times y is greater than or equal to six.

入試に出た数式の英語発音 7

北見工業大学

$$n! \left(\sum_{k=n+1}^{\infty} \frac{1}{k!} \right) < \frac{2}{n+1}$$

n factorial times the sum from k equals n plus one to infinity of one over k factorial is less than two over n plus one.

入試に出た数式の英語発音 8

北見工業大学

$$g(x) = (x - w_1)(x - w_2)\cdots(x - w_n),$$

$$w_k = \cos\frac{2\pi k}{n} + i\sin\frac{2\pi k}{n}$$
$$(k = 1, \cdots, n)$$

g of x equals x minus w sub one times x minus w sub two times, and so on, x minus w sub n, where w sub k equals cosine two pi k over n plus i sine two pi k over n, for k equal to one up to n.

入試に出た数式の英語発音　　9

小樽商科大学

$$f_1(x) = x + 1,$$
$$x^2 f_{n+1}(x) = x^3 + x^2 + \int_0^x t f_n(t) dt$$
$$(n = 1, 2, 3, \cdots)$$

f sub one of x equals x plus one, x squared f sub n plus one of x equals x cubed plus x squared plus the integral from zero to x of t f sub n of t d t, for n taking the values one, two, three, and so on.

入試に出た数式の英語発音 10

小樽商科大学

$$\frac{xe^{-x}}{\log 2} < e^{-\frac{x}{2}}$$

x times e to the minus x over log two is less than e to the minus one half x.

入試に出た数式の英語発音　11

弘前大学

$$\left(\frac{1}{3}\right)^{1-2x} > \frac{1}{\sqrt{27}},$$
$$3^{x+1} + \left(\frac{1}{3}\right)^{x-1} < 10$$

One third to the one minus two x is greater than one over the square root of twenty seven, three to the x plus one plus one third to the x minus one is less than ten.

入試に出た数式の英語発音 12

弘前大学

$$f(x) = 5x - \left|3x + \frac{2}{3} - \log 8\right| - 2\log x$$
$$(x > 0)$$

f of x equals five x minus the absolute value of three x plus two thirds minus log eight minus two log x, for x positive.

入試に出た数式の英語発音 13

岩手大学

$$y = \sin^3 x + \cos^3 x + \sin x \cos x - \sin x - \cos x$$
$$(0° \leq x \leq 180°)$$

y equals sine cubed x plus cosine cubed x plus sine x cosine x minus sine x minus cosine x, [for x between zero degrees and one hundred eighty degrees inclusive / as x varies down to and including zero degrees and up to and including one hundred eighty degrees][1].

---

[1] [……／……] はスラッシュ／の前部と後部に書かれた英語発音のどちらもよく使うという意味です。

入試に出た数式の英語発音　14

岩手大学

$$\frac{1}{5} - \frac{1}{n+1} < \sum_{k=5}^{n} \frac{1}{k^2} < \frac{1}{4} - \frac{1}{n}$$

One fifth minus one over n plus one is less than the sum from k equals five to n of one over k squared is less than one fourth minus one over n.

入試に出た数式の英語発音 15
東北大学

$$I_m = \lim_{t \to \infty} \int_0^t x^m e^{-x} dx$$
$$(m = 0, 1, 2, \cdots)$$

Upper case I sub m equals the limit as t tends to infinity of the integral from zero to t of x to the m times e to the minus x d x, for m taking the values zero, one, two, and so on.

東北大学

$$\lim_{n\to\infty} \frac{1}{n} \left(\frac{1}{\sqrt{a_1}} + \frac{1}{\sqrt{a_2}} + \cdots + \frac{1}{\sqrt{a_n}} \right)$$

The limit as n tends to infinity of one over n times one over the square root of a sub one plus one over the square root of a sub two plus, and so on, plus one over the square root of a sub n.

入試に出た数式の英語発音 17

秋田大学

$$\overrightarrow{OP} = s\overrightarrow{OA} + t\overrightarrow{OB} + (1-s-t)\overrightarrow{OC}$$

The vector OP equals s times the vector OA plus t times the vector OB plus one minus s minus t times the vector OC.

入試に出た数式の英語発音　18
宮城教育大学

$$T = \sum_{n=0}^{\infty} |\alpha_{n+1} - \alpha_n|$$

Capital T equals the sum from n equals zero to infinity of the absolute value of alpha sub n plus one minus alpha sub n.

山形大学

$$\int_{\frac{1}{\sqrt{2}}}^{\frac{\sqrt{3}}{2}} \frac{1-2x^2}{x(1-x^2)} dx$$

The integral from one over root two to root three over two of one minus two x squared over x times one minus x squared $d\,x$.

入試に出た数式の英語発音 20

山形大学

$$f_0(x) = \sin x$$
$$f_{n+1}(x) = \sin x + \left(\int_0^{\frac{\pi}{2}} f_n(t) \sin t\, dt\right) \cos x$$
$$(n = 0, 1, 2, \cdots)$$

f sub zero of x equals sine x, f sub n plus one of x equals sine x plus the integral from zero to one half pi of f sub n of t sine t d t times cosine x, for n equal to zero, one, two, and so on.

茨城大学

$$f(x) = \int_{x}^{2x} \left(\frac{3}{7}t^2 - 4t + 9\right) dt$$

f of x equals the integral from x to two x of three sevenths t squared minus four t plus nine $d\,t$.

入試に出た数式の英語発音　22

筑波大学

$$a_n = \left(1 + \frac{1}{n}f\left(\frac{1}{n}\right)\right)\left(1 + \frac{1}{n}f\left(\frac{2}{n}\right)\right)$$
$$\cdots \left(1 + \frac{1}{n}f\left(\frac{n}{n}\right)\right)$$

a sub n equals one plus one over n f of one over n times one plus one over n f of two over n times, and so on, times one plus one over n f of n over n.

筑波大学

$$\frac{1}{n}\sum_{i=1}^{n} f(x_i) \leq \int_0^1 f(x)dx$$
$$\leq \frac{1}{n}\sum_{i=0}^{n-1} f(x_i)$$

One over n times the sum from i equals one to n of f of x sub i is less than or equal to the integral from zero to one of f of x d x, which is less than or equal to one over n times the sum from i equals zero to n minus one of f of x sub i.

入試に出た数式の英語発音　24
図書館情報大学

$$\alpha = \angle APN,$$
$$\beta = \angle BPC,$$
$$\gamma = \angle APB - (\angle APN + \angle BPC)$$
$$= \angle APB - (\alpha + \beta)$$

Alpha equals the angle APN, beta equals the angle BPC, gamma equals the angle APB minus (parentheses) the angle APN plus the angle BPC, equals the angle APB minus (parentheses) alpha plus beta.

入試に出た数式の英語発音 25

宇都宮大学

$$y' = \frac{dy}{dx}$$
$$= \lim_{h \to 0} \frac{f(x+h) - f(x)}{h}$$
$$\lim_{h \to 0} g(x+h)$$
$$+ f(x) \lim_{h \to 0} \frac{g(x+h) - g(x)}{h}$$

y prime equals $d\, y\, d\, x$ equals the limit as h tends to zero of f of x plus h minus f of x over h times the limit as h tends to zero of g of x plus h plus f of x times the limit as h tends to zero of g of x plus h minus g of x over h.

群馬大学

$$f(x) = \frac{x}{x^2 - ax + a - 1} \quad (x > 0)$$

f of x equals x over x squared minus a x plus a minus one, for x positive.

入試に出た数式の英語発音 27

埼玉大学

$$A = \begin{pmatrix} a & a \\ a & a \end{pmatrix},$$
$$X_n = E + A + A^2 + \cdots + A^n$$
$$= \begin{pmatrix} x_{11}^{(n)} & x_{12}^{(n)} \\ x_{21}^{(n)} & x_{22}^{(n)} \end{pmatrix}$$

A equals the two by two matrix a, a, a, a, X sub n equals E plus A plus A squared plus, and so on, plus A to the n equals the two by two matrix x sub one one super parentheses n, x sub one two super parentheses n, x sub two one super parentheses n, x sub two two super parentheses n.

入試に出た数式の英語発音　28

埼玉大学

$$|\vec{a}| = |\vec{b}| = |\vec{c}| = 1,$$
$$\vec{a} \cdot \vec{b} = \vec{b} \cdot \vec{c} = \vec{c} \cdot \vec{a} = 0$$

The norm of the vector a equals the norm of the vector b equals the norm of the vector c equals one, the scalar product of a and b equals the scalar product of b and c equals the scalar product of c and a equals zero.

入試に出た数式の英語発音 29

千葉大学

$$\begin{pmatrix} \sum_{n=1}^{\infty} a_n & \sum_{n=1}^{\infty} b_n \\ \sum_{n=1}^{\infty} c_n & \sum_{n=1}^{\infty} d_n \end{pmatrix}$$

The two by two matrix with the one one element given by the sum from n equals one to infinity of a sub n, the one two element given by the sum from n equals one to infinity of b sub n, the two one element given by the sum from n equals one to infinity of c sub n, and the two two element given by the sum from n equals one to infinity of d sub n.

千葉大学

$$|\overrightarrow{PA}| : |\overrightarrow{PB}| : |\overrightarrow{PC}| = \sqrt{7} : \sqrt{6} : \sqrt{5}$$

The length of the vector PA is to that of the vector PB to that of the vector PC as root seven is to root six to root five.
[The lengths of the vectors PA, PB, and PC are in proprtion to root seven, root six, root five.]

入試に出た数式の英語発音 31

東京大学

$$f(x) = \frac{a}{2\pi}\int_0^{2\pi} \sin(x+y)f(y)dy \\ + \frac{b}{2\pi}\int_0^{2\pi} \cos(x-y)f(y)dy \\ + \sin x + \cos x$$

f of x equals a over two pi times the integral from zero to two pi of sine x plus y times f of y d y plus b over two pi times the integral from zero to two pi of cosine x minus y times f of y d y plus sine x plus cosine x.

入試に出た数式の英語発音 32

東京大学

$$n - \sum_{k=2}^{n} \frac{k}{\sqrt{k^2 - 1}} \geq \frac{i}{10}$$

n minus the sum from k equals two to n of k over the square root of k squared minus one is greater than or equal to i over ten.

第 II 部 実地訓練

入試に出た数式の英語発音　33

東京大学

$$\{x | k \leq x \leq k+1, P(x) \in I\},$$
$$P(x) = (\cos 2\pi f(x), \sin 2\pi f(x))$$

The set of elements x such that x is between k and k plus one inclusive, and capital P of x belongs to the set capital I, where capital P of x equals the pair cosine two pi f of x, sine two pi f of x.

入試に出た数式の英語発音 34
東京学芸大学

$$\left(\vec{a}\cdot\vec{b}\right)^2 = \frac{1}{2}\left(\vec{a}\cdot\vec{c}+1\right)$$

The scalar product of the vectors a and b squared equals one half times the scalar product of the vectors a and c plus one.

入試に出た数式の英語発音 35

東京工業大学

$$S(a,t) = \int_0^a \left| e^{-x} - \frac{1}{t} \right| dx$$

Capital S of a t equals the integral from zero to a of the absolute value of e to the minus x minus one over t d x.

入試に出た数式の英語発音　36
東京商船大学

$$\int_p^{p+2h} f(x)dx = \frac{h}{3}\left(y_0 + 4y_1 + y_2\right)$$

The integral from p to p plus two h of f of x d x equals h over three times y sub zero plus four y sub one plus y sub two.

入試に出た数式の英語発音　37
東京水産大学

$$a_{n+1} = (n+1)\left[\frac{a_n}{n+1}\right]$$
$$(n = 1, 2, 3, \cdots)$$

a sub n plus one equals n plus one times a sub n over n plus one, for n taking the values one, two, three, and so on.

入試に出た数式の英語発音 38

お茶の水女子大学

$$(1 - b_1)(1 - b_2) \cdots (1 - b_n) \geq 1 - \sum_{i=1}^{n} b_i$$

One minus b sub one times one minus b sub two times, and so on, times one minus b sub n is greater than or equal to one minus the sum from i equals one to n of b sub i.

入試に出た数式の英語発音　39
お茶の水女子大学

$$S(2, n) \subseteq S(m, n),$$
$$S(2, n) \cap S(m, n) = \emptyset$$

Capital S of two n is included in or equal to capital S of $m\ n$, the intersection of capital S of two n and capital S of $m\ n$ equals the empty set.

電気通信大学

$$\frac{1}{2} - \frac{1}{3}h \leq \int_0^1 \frac{x}{1+hx}dx \leq \frac{1}{2} + \frac{2}{3}|h|$$

The integral from zero to one of x over one plus h x d x is between one half minus one third h and one half plus two thirds the absolute value of h, inclusive.

入試に出た数式の英語発音　41
電気通信大学

$$f(x) = \begin{cases} \frac{\log(1+x)}{x} & (0 < |x| < 1) \\ 1 & (x = 0) \end{cases}$$

f of x equals log of one plus x over x provided that the absolute value of x lies between zero and one, and equals one for x equal to zero.

入試に出た数式の英語発音　42

一橋大学

$$|\overrightarrow{OA}| = 1,$$
$$\overrightarrow{OA} \perp \overrightarrow{OP}, \overrightarrow{OP} \perp \overrightarrow{OQ}, \overrightarrow{OA} \perp \overrightarrow{OQ},$$
$$\angle PAQ = 30°$$

The length of the vector OA equals one, the vector OA is perpendicular to the vector OP, the vector OP is perpendicular to the vector OQ, the vector OA is perpendicular to the vector OQ, and the angle PAQ equals thirty degrees.

入試に出た数式の英語発音 43
横浜国立大学

$$\lim_{\theta \to \frac{3}{4}\pi + 0} f(\theta) \sin\left(\theta - \frac{3}{4}\pi\right)$$

The limit as theta tends to three fourths pi from above of f of theta times sine of theta minus three fourths pi.

横浜国立大学

$$b_n = 1 + \sum_{k=1}^{n} {}_nC_k a_k \ (n = 1, 2, 3, \cdots)$$

b sub n equals one plus the sum from k equals one to n of the number of combinations of n elements taken k at a time times a sub k, for n taking the values one, two, three, and so on.

新潟大学

$$\int_0^x \frac{1}{1-t^2} dt$$
$$= x + \frac{x^3}{3} + \frac{x^5}{5} + \cdots$$
$$+ \frac{x^{2n-1}}{2n-1} + \int_0^x \frac{t^{2n}}{1-t^2} dt$$

The integral from zero to x of one over one minus t squared $d\,t$ equals x plus x cubed over three plus x to the fifth over five plus, and so on, plus x to the two n minus one over two n minus one plus the integral from zero to x of t to the two n over one minus t squared $d\,t$.

入試に出た数式の英語発音　46

富山大学

$$a_n = \int_0^\pi x e^{-\frac{x}{n}} \sin x \, dx$$
$$(n = 1, 2, 3, \cdots)$$

a sub n equals the integral from zero to pi of x e to the minus x over n sine x d x, for n taking the values one, two, three, and so on.

富山医科薬科大学

$$\int_{n+1}^{2n+2} \frac{1}{x}dx < S_{2n+1} - S_n < \int_{n}^{2n+1} \frac{1}{x}dx$$

The integral from n plus one to two n plus two of one over x d x is less than capital S sub two n plus one minus capital S sub n, which is less than the integral from n to two n plus one of one over x d x.

入試に出た数式の英語発音　48
富山医科薬科大学

$$A = \{(x, y) | x \geq 0 \text{ かつ } y \geq 0$$
$$\text{かつ } x^2 + y^2 \leq 1\}$$

Capital A equals the set of pairs x y such that x is greater than or equal to zero, and y is greater than or equal to zero, and x squared plus y squared is less than or equal to one.

金沢大学

$$E + \frac{1}{2}A + \left(\frac{1}{2}A\right)^2 + \cdots + \left(\frac{1}{2}A\right)^{2001},$$
$$A = \begin{pmatrix} 1 & -5 \\ 1 & -1 \end{pmatrix}, E = \begin{pmatrix} 1 & 0 \\ 0 & 1 \end{pmatrix}$$

Capital E plus one half capital A plus one half capital A squared plus, and so on, plus one half capital A to the two thousand one, where capital A equals the two by two matrix, one, minus five, one, minus one, and capital E equals the two by two matrix, one, zero, zero, one.

入試に出た数式の英語発音 50

金沢大学

$$f(x) = \begin{cases} qx^2 + px + 2 & (0 \leq x < \alpha) \\ -x^2 + 1 & (\alpha \leq x \leq 1) \end{cases}$$

f of x equals q x squared plus p x plus two, for x greater than or equal to zero and less than alpha, and equals minus x squared plus one, for x greater than or equal to alpha and less than or equal to one.

入試に出た数式の英語発音　51

金沢大学

$$z_1 = 0,$$
$$z_{n+1} = -\frac{2}{3} \times \frac{4z_n + 5}{z_n + 1},$$
$$(n = 1, 2, 3, \cdots)$$

z sub one equals zero, z sub n plus one equals minus two thirds times four z sub n plus five over z sub n plus one, for n taking the values one, two, three, and so on.

入試に出た数式の英語発音　52

金沢大学

$$\overrightarrow{OP} = \cos\theta\vec{u} + \sqrt{3}\sin\theta\vec{v},$$
$$(0 \leq \theta < 2\pi)$$

The vector OP equals cosine theta times the vector u plus root three sine theta times the vector v, for theta greater than or equal to zero and less than two pi.

入試に出た数式の英語発音 53

福井大学

$$n(A) = 65, n(B) = 40,$$
$$n(A \cap B) = 14,$$
$$n(A \cap C) = 11, n(A \cup C) = 78,$$
$$n(B \cup C) = 55, n(A \cup B \cup C) = 99$$

n of A equals sixty five, n of B equals fourty, n of the intersection of A and B equals fourteen, n of the intersection of A and C equals eleven, n of the union of A and C equals seventy eight, n of the union of B and C equals fifty five, n of the union of A, B and C equals ninety nine.

入試に出た数式の英語発音　54

福井医科大学

$$\frac{\sin\theta_1 + \sin\theta_2 + \cdots + \sin\theta_n}{n}$$
$$\leq \sin\left(\frac{\theta_1 + \theta_2 + \cdots + \theta_n}{n}\right)$$

Sine theta sub one plus sine theta sub two plus, and so on, plus sine theta sub n over n is less than or equal to sine of theta sub one plus theta sub two plus, and so on, plus theta sub n over n.

第Ⅱ部 実地訓練

入試に出た数式の英語発音　55

山梨医科大学

$$\sum_{r=0}^{n} {}_nC_r e^{r\alpha+n\beta} = 1$$

The sum from r equals zero to n of the number of combinations of n elements taken r at a time times e to the r alpha plus n beta equals one.

入試に出た数式の英語発音 56
山梨医科大学

$$\left|\frac{g_n(x)\cos mx}{f(x)}\right| \leq \frac{2(1+|a|)}{(1-|a|)^2}|a|^{n+1}$$

The absolute value of g sub n of x cosine $m\,x$ over f of x is less than or equal to two times one plus the absolute value of a over one minus the absolute value of a squared times the absolute value of a to the n plus one.

入試に出た数式の英語発音　57
山梨大学

$$1^2 + 2^2 + \cdots + n^2$$
$$= \frac{n(n+1)(2n+1)}{6}$$

One squared plus two squared plus, and so on, plus n squared equals n times n plus one times two n plus one over six.

入試に出た数式の英語発音　58

山梨大学

$$\overline{\left(\frac{1}{z}\right)} = z + m\alpha + m\beta$$

The complex conjugate of one over z equals z plus m alpha plus m beta.

信州大学

$$f_n(x) = (\underbrace{f \circ \cdots \circ f}_{n})(x)$$

f sub n of x equals the n-fold composition of f of x.

入試に出た数式の英語発音　60
信州大学

$$(y - ax - b)(y - x + 2)(3y - x) > 0$$

y minus a x minus b times y minus x plus two times three y minus x is [positive / greater than zero].

入試に出た数式の英語発音 61

岐阜大学

$$\begin{cases} (y - x^3 + x^2 + x - 1)(y - 1 + x^2) \\ \qquad \leq 0 \\ -1 \leq x \leq 1 \end{cases}$$

y minus x cubed plus x squared plus x minus one times y minus one plus x squared is less than or equal to zero, and x is greater than or equal to minus one, and less than or equal to one.

入試に出た数式の英語発音　62

静岡大学

$$(x+1)\{f(x) - \log(x+1) + 1\}$$
$$= \int_0^x f(t)dt \quad (x > -1)$$

x plus one times f of x minus log of x plus one plus one equals the integral from zero to x of f of t d t, for x greater than minus one.

入試に出た数式の英語発音

浜松医科大学

$$\int_{\frac{\pi}{2}}^{\pi} \frac{x \sin x}{1+|\cos x|} dx = \pi \int_{0}^{\frac{\pi}{2}} \frac{\sin t}{1+\cos t} dt - \int_{0}^{\frac{\pi}{2}} \frac{t \sin t}{1+\cos t} dt$$

The integral from one half pi to pi of x sine x over one plus the absolute value of cosine x d x equals pi times the integral from zero to one half pi of sine t over one plus cosine t d t minus the integral from zero to one half pi of t sine t over one plus cosine t d t.

名古屋大学

$$\log(\log q) - \log(\log p) < \frac{q-p}{e}$$
$$(e \leq p < q)$$

Log log q minus log log p is less than q minus p over e, for p greater than or equal to e and less than q.

名古屋大学

$$f(x) = -|2x - 1| + 1 \quad (0 \leq x \leq 1)$$
$$g(x) = -|2f(x) - 1| + 1$$
$$(0 \leq x \leq 1)$$

f of x equals minus the absolute value of two x minus one plus one, for x between zero and one inclusive, g of x equals minus the absolute value of two f of x minus one plus one, for x between zero and one inclusive.

入試に出た数式の英語発音 66

名古屋大学

$$\begin{pmatrix} a_{n+2} \\ a_{n+1} \end{pmatrix} = \begin{pmatrix} p+1 & -2p \\ -2q & q+2 \end{pmatrix} \begin{pmatrix} a_{n+1} \\ a_n \end{pmatrix}$$

$$(n = 1, 2, 3, \cdots)$$

The two by one matrix, a sub n plus two, a sub n plus one equals the two by two matrix, p plus one, minus two p, minus two q, q plus two, times the two by one matrix, a sub n plus one, a sub n, for n taking the values one, two, three, and so on.

入試に出た数式の英語発音　67

愛知教育大学

$$|z-\alpha|^2 + |z-\beta|^2 = |z|^2$$

The absolute value of z minus alpha squared plus the absolute value of z minus beta squared equals the absolute value of z squared.

入試に出た数式の英語発音 68
名古屋工業大学

$$z = \cos\theta + i\sin\theta,$$
$$F_2(z) = \frac{1}{4}(1 + z + z^2 + z^3),$$
$$|F_2(z)| = \left|\cos\theta \cos\frac{\theta}{2}\right|$$

z equals cosine theta plus i sine theta, F sub two of z equals one fourth times one plus z plus z squared plus z cubed, and the absolute value of F sub two of z equals the absolute value of cosine theta cosine one half theta.

入試に出た数式の英語発音

豊橋技術科学大学

$$\begin{pmatrix} 1 & k \\ 0 & 2 \end{pmatrix}^n = \begin{pmatrix} 1 & (2^n - 1) \cdot k \\ 0 & 2^n \end{pmatrix}$$

The two by two matrix, one, k, zero, two, to the n equals the two by two matrix, one, two to the n minus one times k, zero, two to the n.

三重大学

$$\log_{10}(M+N) \geq \frac{1}{2}(\log_{10} M + \log_{10} N) + \log_{10} 2$$

Log base ten of M plus N is greater than or equal to one half times parentheses log base ten of M plus log base ten of N, close parentheses, plus log base ten of two.

入試に出た数式の英語発音　71
三重大学

$$\begin{pmatrix} 5 & 2 \\ -8 & -3 \end{pmatrix}^p + \begin{pmatrix} 7 & 3 \\ -12 & -5 \end{pmatrix}^q = \begin{pmatrix} 38 & 18 \\ -72 & -34 \end{pmatrix}$$

The two by two matrix, five, two, minus eight, minus three to the p plus the two by two matrix, seven, three, minus twelve, minus five to the q equals the two by two matrix, thirty eight, eighteen, minus seventy two, minus thirty four.

入試に出た数式の英語発音 72

滋賀医科大学

$$U = \left\{ \begin{pmatrix} 1 & b \\ 0 & d \end{pmatrix} \middle| b, d \text{ real}, d \neq 0 \right\}$$

U equals the set of two by two matrices, one, b, zero, d such that b d are real numbers and d does not vanish.

京都大学

$$\lim_{n\to\infty} \int_0^{n\pi} e^{-x} \left|\sin nx\right| dx$$

The limit as n tends to infinity of the integral from zero to n pi of e to the minus x times the absolute value of sine n x d x.

入試に出た数式の英語発音　　74

京都大学

$$|\vec{u}| = 1,$$
$$|\vec{u} + 3\vec{v}| = 1,$$
$$|2\vec{u} + \vec{v}| = \sqrt{2}$$

The norm of the vector u is one, the norm of the vector u plus three times the vector v is one, and the norm of two times the vector u plus the vector v is the square root of two.

入試に出た数式の英語発音 75
京都大学

$$E_n = 3 - \left(\frac{a+b}{N}\right)^n - \left(\frac{b+c}{N}\right)^n - \left(\frac{c+a}{N}\right)^n$$

Upper case E sub n equals three minus a plus b over upper case N to the n minus b plus c over upper case N to the n minus c plus a over upper case N to the n.

京都教育大学

$$_nC_k = \frac{n!}{(n-k)!k!}$$

The number of combinations of n elements taken k at a time equals n factorial over n minus k factorial times k factorial.

京都教育大学

$$L_1 = \int_0^{t_0} \sqrt{\left(\frac{dx}{dt}\right)^2 + \left(\frac{dy}{dt}\right)^2} dt$$

Upper case L sub one equals the integral from zero to t sub zero of the square root of $d\,x\,d\,t$ squared plus $d\,y\,d\,t$ squared $d\,t$.

大阪大学

$$\sum_{n=1}^{\infty} a_n = \frac{1}{2},$$
$$b_n = (1-a_1)(1-a_2)\cdots(1-a_n),$$
$$c_n = 1 - (a_1 + a_2 + \cdots + a_n)$$

The sum from n equals one to infinity of a sub n equals one half, b sub n equals one minus a sub one times one minus a sub two times, and so on, times one minus a sub n, c sub n equals one minus parentheses a sub one plus a sub two plus, and so on, plus a sub n close parentheses.

大阪教育大学

$$\theta_1 = \frac{\pi}{2},$$
$$\sin\theta_{n+1} = \frac{\sqrt{1-\sqrt{1-\sin^2\theta_n}}}{\sqrt{2}}$$
$$(n = 1, 2, 3, \cdots)$$

Theta sub one equals one half pi, sine theta sub n plus one equals the square root of one minus the square root of one minus sine squared theta sub n over the square root of two, for n taking the values one, two, three, and so on.

神戸大学

$$|(1+i)^n| = |(1+\sqrt{3}i)^m|$$

The absolute value of one plus i to the n equals the absolute value of one plus root three i to the m.

神戸商船大学

$$a^2b^2 + b^2c^2 + c^2a^2 \geq 3(abc)^{\frac{4}{3}}$$

a squared b squared plus b squared c squared plus c squared a squared is greater than or equal to three a b c to the four thirds.

入試に出た数式の英語発音 82
奈良女子大学

$$\lim_{h \to +0} \frac{F(p+h) - F(p)}{h}$$
$$= \lim_{h \to -0} \frac{F(p+h) - F(p)}{h}$$

The limit as h tends to zero from above of upper case F of p plus h minus upper case F of p over h equals the limit as h tends to zero from below of upper case F of p plus h minus upper case F of p over h.

入試に出た数式の英語発音 83

和歌山大学

$$f(x) = x^3 e^x - 14e^{-x} \\ + \int_0^x 2e^{2t-x}(-6t^2 + 8t - 7)dt$$

f of x equals x cubed e to the x minus fourteen e to the minus x plus the integral from zero to x of two e to the two t minus x times minus six t squared plus eight t minus seven $d\,t$.

入試に出た数式の英語発音　84

島根医科大学

$$N = \sum_{i=1}^{m} \left(A_i \times 100^{i-1}\right)$$

Upper case N equals the sum from i equals one to m of upper case A sub i times one hundred to the i minus one.

広島大学

$$f(x) = x^2 - \int_0^x (x-t) f'(t) dt$$

f of x equals x squared minus the integral from zero to x of x minus t times f prime of t $d\,t$.

入試に出た数式の英語発音　86

広島大学

$$\frac{x^2}{a^2} + a^2\left(y - \frac{1}{a}\right)^2 = 1$$

x squared over a squared plus a squared times y minus one over a squared equals one.

入試に出た数式の英語発音　87

山口大学

$$f(x) = |x - i_1| + |x - i_2| + \cdots + |x - i_n|$$

f of x equals the absolute value of x minus i sub one plus the absolute value of x minus i sub two plus, and so on, plus the absolute value of x minus i sub n.

入試に出た数式の英語発音　88

徳島大学

$$X^n = E + nA + \frac{n(n-1)}{2}A^2$$

Capital X to the n equals capital E plus n times capital A plus n times n minus one over two times capital A squared.

入試に出た数式の英語発音 89

徳島大学

$$\int_\alpha^\beta \frac{x}{1+x^2}dx = \log \beta$$
$$\left(0 < \alpha < \beta, \frac{\alpha}{1+\alpha^2} = \frac{\beta}{1+\beta^2}\right)$$

The integral from alpha to beta of x over one plus x squared dx equals log beta, where alpha is greater than zero and less than beta, and alpha over one plus alpha squared equals beta over one plus beta squared.

入試に出た数式の英語発音　　90

徳島大学

$$f_1(x) = (x^2 - 10x + 30)e^x,$$
$$f_{n+1}(x) = f'_n(x) \ (n = 1, 2, 3, \cdots)$$

f sub one of x equals parentheses x squared minus ten x plus thirty close parentheses times e to the x, f sub n plus one of x equals f sub n prime of x, for n taking the values one, two, three, and so on.

入試に出た数式の英語発音

鳴門教育大学

$$\sqrt[n]{a} < \sqrt[n]{b} \quad (0 < a < b)$$

The n-th root of a is less than the n-th root of b, for a greater than zero and less than b.

入試に出た数式の英語発音　92
香川医科大学

$$|\alpha + \beta + \gamma| = \left|\frac{1}{\alpha} + \frac{1}{\beta} + \frac{1}{\gamma}\right|$$

The absolute value of alpha plus beta plus gamma equals the absolute value of one over alpha plus one over beta plus one over gamma.

入試に出た数式の英語発音

高知大学

$$T_{2n} = \frac{2S_n T_n}{S_n + T_n},$$
$$S_{2n} = \sqrt{S_n T_{2n}}$$

Capital T sub two n equals two capital S sub n capital T sub n over capital S sub n plus capital T sub n, capital S sub two n equals the square root of capital S sub n capital T sub two n.

入試に出た数式の英語発音　94

高知医科大学

$$y = \frac{\left(x + \frac{1}{x}\right)^3 - \left(x^3 + \frac{1}{x^3}\right) - 4}{x^2 + \frac{1}{x^2} + 3}$$
$$(x > 0)$$

y equals the cube of x plus one over x minus x cubed plus one over x cubed minus four, over x squared plus one over x squared plus three, for x positive.

入試に出た数式の英語発音

高知医科大学

$$A_n = \sum_{i=1}^{n} x_i \log \sqrt[n]{x_i}$$

Upper case A sub n equals the sum from i equals one to n of x sub i log of the n-th root of x sub i.

入試に出た数式の英語発音　96

愛媛大学

$$\sin\alpha + \sin\beta = 2\sin\frac{\alpha+\beta}{2}\cos\frac{\alpha-\beta}{2}$$

Sine alpha plus sine beta equals twice sine alpha plus beta over two times cosine alpha minus beta over two.

九州工業大学

$$f(t) = \frac{2e}{t} + \frac{2t}{e} - 5,$$
$$g(t) = 4\log t$$
$$\left(\frac{e}{2} \leq t \leq 2e\right)$$

f of t equals two e over t plus two t over e minus five, g of t equals four log t, for t between e over two and two e inclusive.

九州工業大学

$$\frac{(x-q)^2}{s^2} + \frac{y^2}{t^2} = 1$$
$$(q>0, s>0, t>0)$$

x minus q squared over s squared plus y squared over t squared equals one, for q, s and t all positive.

福岡教育大学

$$\log_{10}\left(\frac{|x-1|}{3}\right) = \log_{\frac{1}{10}}|x-5|$$

Log base ten of the absolute value of x minus one over three equals log base one tenth of the absolute value of x minus five.

入試に出た数式の英語発音　　100
九州大学

$$\lim_{n\to\infty} n^2 f\left(\frac{1}{n}\right) = 0$$

The limit as n tends to infinity of n squared times f of one over n equals zero.

九州大学

$$\sum_{i=1}^{n} \frac{1}{x_i} \geq \frac{n^2}{\sum_{i=1}^{n} x_i}$$

The sum from i equals one to n of one over x sub i is greater than or equal to n squared over the sum from i equals one to n of x sub i.

入試に出た数式の英語発音　　102
九州芸術工科大学

$$\begin{cases} x(\theta) = \displaystyle\int_0^\theta \cos^3 t\, dt \\ y(\theta) = \theta \cos\theta \end{cases} \left(0 \leq \theta \leq \tfrac{\pi}{2}\right)$$

x of theta equals the integral from zero to theta of cosine cubed t d t, y of theta equals theta cosine theta, for theta between zero and pi over two inclusive.

宮崎大学

$$f_n(x) = x^2 - \frac{n}{n+1}x^{2+\frac{2}{n}}$$

f sub n of x equals x squared minus n over n plus one times x to the two plus two over n.

鹿児島大学

$$\lim_{n\to\infty}\int_{\frac{1}{n}}^{\alpha}\left(2x^2\log x - kx^2 + k\right)dx$$

The limit as n goes to infinity of the integral from one over n to alpha of two x squared log x minus k x squared plus k d x.

鹿児島大学

$$\begin{pmatrix} 2 & 1 \\ 0 & 2 \end{pmatrix}^n = \begin{pmatrix} 2^n & n2^{n-1} \\ 0 & 2^n \end{pmatrix}$$

The two by two matrix, two, one, zero, two, to the n equals the two by two matrix, two to the n, n times two to the n minus one, zero, two to the n.

第III部

実 践

大学で学ぶ方程式

再び大学受験生の皆さんへ

これまでは Let's speak Mathematics! の基礎訓練と実地訓練ということで，(大学で習うものもありましたが) 主として高校で習う範囲の数学公式や，2001 年に行われた大学入試で出題された数式を英語で読んできました。

数学の内容そのものでしたら，高校生の皆さんが大学や大学院でしか出てこないような高度な数学や物理学の方程式などをすぐに理解することは不可能に近いことかもしれません。しかし，方程式をサラリと英語で読むだけならば，何も大学生になるまで待つ必要はありません。これまでに皆さんが見てきた Let's speak Mathematics! の内容を応用するだけで，ノーベル賞学者たちでさえ頭を悩ませてきた難解な方程式だって，ほんの少しがんばるだけでクリアーできるのです。

さあ，受験勉強に飽きた高校生の皆さん。ここは，気分転換にちょっとだけ背伸びしてみませんか？

もちろん，同じ英語で読むなら少しくらいは方程式の意味するところがわかるほうが面白いはずです。だから，ほんのちょっとした解説もつけておきましょう。

第 III 部　実　践＝大学で学ぶ方程式

相対性理論の方程式の英語発音

●ローレンツ変換
Lorentz transformation

$$x' = \frac{x - vt}{\sqrt{1 - \frac{v^2}{c^2}}}$$

$$y' = y$$

$$z' = z$$

$$ct' = \frac{ct - \frac{v}{c}x}{\sqrt{1 - \frac{v^2}{c^2}}}$$

・・

x prime equals x minus v t over the square root of one minus v squared over c squared, y prime equals y, z prime equals z, c t prime equals c t minus v over c times x over the square root of one minus v squared over c squared.

・・

空間座標 x, y, z と時間座標 t とからなる 4 次元の**時空座標系**[1] x, y, z, ct（c は真空中での光の速さを表す定数）を持つ慣性系 inertial system に対して x 軸方向に速さ v で**等速直線運動** uniform motion をする別の慣性系での時空座標系を x', y', z', ct' としたとき，両者の間に成り立つ変換公式がローレンツ変換の公式です。

・・

---
[1] 時空座標系 system of space-time coordinates

相対性理論の方程式の英語発音

●アインシュタインの公式
Einstein's formula

$$E = mc^2$$

・・・・・・・・・・・・・・・・・・・・・・・・・・・・・・

E equals *m* *c* squared.

・・・・・・・・・・・・・・・・・・・・・・・・・・・・・・

ネクタイのデザインになっているほど有名な公式ですが，その意味は静止して存在する**質量**[2] m の**物体** matter が消滅するとき，質量 m に真空中の光の速さ c の2乗をかけた値の**エネルギー** energy が発生すると理解することもでき，一部では**原爆** atomic bomb の原理だといわれています。そのため，原爆開発とは直接に関わりのなかったアインシュタインを原爆の父だと誤解している人もいます。**アインシュタイン** Einstein の**特殊相対性理論** special theory of relativity の重要な公式となっています。

・・・・・・・・・・・・・・・・・・・・・・・・・・・・・・

---

[2] 質量 mass

第 III 部　実　践＝大学で学ぶ方程式

●時空計量の式
space-time metric form

$$c^2 d\tau^2 = \sum_{\mu,\nu=1}^{4} g_{\mu\nu}(x) dx^\mu dx^\nu$$
$$= g_{\mu\nu}(x) dx^\mu dx^\nu$$

....

c squared d tau squared equals the sum from mu and nu equal one to four of g sub mu nu of x d x super mu d x super nu, equals g sub mu nu of x d x super mu d x super nu.

....

重力 gravity の存在までも考慮した**相対性理論** theory of relativity は**一般相対性理論** general theory of relativity と呼ばれ，やはりアインシュタインによって見いだされました。そこでは**時空座標** space-time coordinates の微小な変化 $(dx, dy, dz, cdt) = (dx^1, dx^2, dx^3, dx^4)$ にともなう時空の長さ（**時空計量** space-time metric という）$cd\tau$ は**重力場** gravity field を表す量（**計量テンソル** metric tensor という）g によって定められます。

....

247

相対性理論の方程式の英語発音

●測地線の方程式
geodesics equation

$$\frac{d^2 x^\mu}{d\tau^2} + \Gamma^\mu_{\nu\sigma} \frac{dx^\nu}{d\tau} \frac{dx^\sigma}{d\tau} = 0$$

....

d squared x super mu d tau squared plus gamma super mu sub nu sigma times $d\,x$ super nu d tau times $d\,x$ super sigma d tau equals zero.

....

アインシュタインの一般相対性理論では，与えられた重力場の下での重力の作用によって**自由落下** free fall する物体は4次元時空の中の最短経路（**測地線** geodesic という）に沿って運動することになりますが，その最短経路を定める方程式がこの測地線の方程式です。

....

第 III 部　実　践＝大学で学ぶ方程式

● アインシュタインの重力場方程式
Einstein's equation of gravity field

$$R^{\mu\nu}(x) - \frac{1}{2}R(x)g^{\mu\nu}(x) = \kappa T^{\mu\nu}(x)$$

・・・・・・・・・・・・・・・・・・・・・・・・・・・・・・・・・・

R super mu nu of x minus one half R of x times g super mu nu of x equals kappa times T super mu nu of x.

・・・・・・・・・・・・・・・・・・・・・・・・・・・・・・・・・・

逆に重力場を表す計量テンソル g は，4次元時空の中での物質の分布エネルギーや運動量 momentum を表すエネルギー・運動量テンソル[3] T が与えられたとき，この重力場の方程式に従って定められます。$R^{\mu\nu}(x)$ は曲率テンソル curvature tensor, $R(x)$ はスカラー曲率 scalar curvature と呼ばれ計量テンソル $g(x)$ の2階微分で書けます。また，κ は重力定数 gravitational constant に関連する定数です。

・・・・・・・・・・・・・・・・・・・・・・・・・・・・・・・・・・

---
[3] エネルギー・運動量テンソル energy-momentum tensor

相対性理論の方程式の英語発音

●シュワルツシルト計量
Schwarzschild's metric

$$c^2 d\tau^2 = \left(1 - \frac{2GM}{c^2 r}\right) c^2 dt^2 - \frac{1}{1 - \frac{2GM}{c^2 r}} dr^2 - r^2(d\theta^2 + \sin^2\theta d\phi^2)$$

・・・・・・・・・・・・・・・・・・・・・・・・・・・・・・・・・・・・・

c squared d tau squared equals parentheses one minus two G M over c squared r close parentheses c squared d t squared minus one over one minus two G M over c squared r times d r squared minus r squared times d theta squared plus sine squared theta d phi squared.

・・・・・・・・・・・・・・・・・・・・・・・・・・・・・・・・・・・・・

重力場のアインシュタイン方程式はとても難しい方程式で，考え出したアインシュタインでさえそれを解いて解を見つけることができませんでした。それを最初に解いてみせたのがシュワルツシルト Schwarzschild で，その計量テンソル g で与えられる時空計量がこの式です。

・・・・・・・・・・・・・・・・・・・・・・・・・・・・・・・・・・・・・

第 III 部　実　践＝大学で学ぶ方程式

古典力学の方程式の英語発音

●ニュートンの運動方程式
Newton's equation of motion

$$m\frac{d^2 x_i}{dt^2} = -\frac{\partial V}{\partial x_i}, \ (i = 1, 2, 3)$$

・・

m times d squared x sub i d t squared equals minus the partial of V with respect to x sub i, for i equals one, two, three.

・・

ニュートンの運動方程式 equation of motion は $F = ma$ として，つまり物体に作用する力 force の大きさはその結果物体に生じた加速度 acceleration に物体の質量 mass をかけた値に等しいという法則として高校の物理 physics でも習いますが，力が位置エネルギー[4] V の勾配 gradient の逆向きで与えられるとして，微分方程式 differential equation の形で書いたものが上の方程式です。

・・

---

[4] 位置エネルギー potential energy

古典力学の方程式の英語発音

●ハミルトンの最小作用原理
Hamilton's principle of least action

$$\delta \int_a^b L\left(q(t), \frac{dq(t)}{dt}, t\right) dt = 0$$

・・・

The variation of the integral from a to b of L of q of t, $d\,q$ of $t\,d\,t$, $t\,d\,t$ equals zero.

・・・

力学の基本原理としてはニュートンの運動方程式よりもこのハミルトンの最小作用原理 principle of least action が用いられます。それは，ニュートンの運動方程式そのものも，さらに**エネルギー保存則** energy conservation law や**運動量保存則** momentum conservation law などの**保存法則** conservation law も最小作用原理から導き出すことができるからです。この最小作用の原理の意味は，**運動エネルギー** kinetic energy と位置エネルギーの差（ラグランジアン Lagrangian という）を**運動経路** path に沿って積分した値は，運動経路を少しだけ変えても（運動経路の**変分** variation という）変わらないということです。

・・・

第 III 部　実　践＝大学で学ぶ方程式

●オイラー-ラグランジュ方程式
Euler-Lagrange equation

$$\frac{d}{dt}\frac{\partial L}{\partial \frac{dq_k}{dt}} - \frac{\partial L}{\partial q_k} = 0, \ (k=1,2,\cdots,n)$$

・・

$d\ d\ t$ the partial of L with respect to $d\ q$ sub $k\ d\ t$ minus the partial of L with respect to q sub k equals zero, for k taking the values one, two, and so on, n.

・・

最小作用原理に従う運動経路が満たす微分方程式がオイラー-ラグランジュ方程式です。運動エネルギーと位置エネルギーの差として与えられるラグランジアン L についてオイラー-ラグランジュ方程式を書き下してみると、それはニュートンの運動方程式に一致します。

・・

古典力学の方程式の英語発音

●ハミルトンの正準方程式
Hamilton's canonical equation

$$\frac{dq_k}{dt} = \frac{\partial H(q,p,t)}{\partial p_k}$$
$$\frac{dp_k}{dt} = -\frac{\partial H(q,p,t)}{\partial q_k}$$

・・・・・・・・・・・・・・・・・・・・・・・・・・・・・・・・

$d\,q$ sub $k\,d\,t$ equals the partial of H of $q,\,p,\,t$ with respect to p sub k, $d\,p$ sub $k\,d\,t$ equals minus the partial of H of $q,\,p,\,t$ with respect to q sub k.

・・・・・・・・・・・・・・・・・・・・・・・・・・・・・・・・

力学の基本法則を与える最小作用原理から導かれるオイラー-ラグランジュ方程式としてのニュートンの運動方程式は位置についての2階の常微分方程式 second order ordinary differential equation ですが, これを位置と運動量についての1階の常微分方程式 first order ordinary differential equation として書き直したものがハミルトンの正準方程式です。

・・・・・・・・・・・・・・・・・・・・・・・・・・・・・・・・

●ハミルトン‐ヤコビ方程式
Hamilton-Jacobi equation

$$\frac{\partial S}{\partial t} + H\left(q, \frac{\partial S}{\partial q}, t\right) = 0$$

・・・

The partial of S with respect to t plus H of q, the partial of S with respect to q, t equals zero.

・・・

最小作用原理に従う運動経路の個々はそれぞれオイラー‐ラグランジュ方程式やハミルトンの正準方程式を満たします。このような個々の運動経路を全て描き出せば,全体としては運動の流れ flow の**場** field のようなものが浮かび上がってきますが,この運動の流れの場が満たすべき方程式がハミルトン‐ヤコビ方程式です。

・・・

電磁気学の方程式の英語発音

●ベクトル場の発散と回転

divergence and rotation of a vector field

$$\operatorname{div} \vec{B} \equiv \vec{\nabla} \cdot \vec{B}$$
$$= \left(\frac{\partial}{\partial x_1}, \frac{\partial}{\partial x_2}, \frac{\partial}{\partial x_3}\right) \cdot (B_1, B_2, B_3)$$
$$= \frac{\partial B_1}{\partial x_1} + \frac{\partial B_2}{\partial x_2} + \frac{\partial B_3}{\partial x_3}$$
$$\operatorname{rot} \vec{B} \equiv \vec{\nabla} \times \vec{B}$$
$$= \left(\frac{\partial}{\partial x_1}, \frac{\partial}{\partial x_2}, \frac{\partial}{\partial x_3}\right) \times (B_1, B_2, B_3)$$
$$= \left(\frac{\partial B_3}{\partial x_2} - \frac{\partial B_2}{\partial x_3}, \frac{\partial B_1}{\partial x_3} - \frac{\partial B_3}{\partial x_1}, \frac{\partial B_2}{\partial x_1} - \frac{\partial B_1}{\partial x_2}\right)$$

....

The divergence of the vector B is defined to be the inner product of the vector nabla and the vector B, equals the inner product of the vector with the components the partial with respect to x sub one, the partial with respect to x sub two, the partial with respect to x sub three and the vector with components B sub one, B sub two, B sub three, equals the partial of B sub one with respect to x sub one plus the partial of B sub two with respect to x sub two plus the partial of B sub three with respect to x sub three. The rotation of the

第 III 部　実　践＝大学で学ぶ方程式

vector B is defined to be the vector product of the vector nabla and the vector B, equals the vector product of the vector with the components the partial with respect to x sub one, the partial with respect to x sub two, the partial with respect to x sub three and the vector with components B sub one, B sub two, B sub three, equals the vector with components the partial of B sub three with respect to x sub two minus the partial of B sub two with respect to x sub three, the partial of B sub one with respect to x sub three minus the partial of B sub three with respect to x sub one, the partial of B sub two with respect to x sub one minus the partial of B sub one with respect to x sub two.

・・・

　空間 space の各点に矢印で表される量が分布したものがベクトル場 vector field です。各点での矢印がその点での流れの向きと大きさを表していると考えたとき，その点から周囲の点に流出していく量と周囲の点から流入してくる量の差は，その点でわき出してくる量の大きさを表し，ベクトル場の**発散** divergence（ダイバージェンス）と呼ばれます。また，その点の周囲の点での流れの向きがその点を 1 周するように回転している流れを与える量がベクトル場の**回転** rotation（ローテーション）です。

・・・

257

電磁気学の方程式の英語発音

●マックスウェル方程式
Maxwell equation

$$\vec{\nabla} \cdot \vec{E} = \rho$$
$$\vec{\nabla} \cdot \vec{B} = 0$$
$$\vec{\nabla} \times \vec{E} = -\frac{1}{c}\frac{\partial \vec{B}}{\partial t}$$
$$\vec{\nabla} \times \vec{B} = \frac{1}{c}\frac{\partial \vec{E}}{\partial t} + \frac{1}{c}\vec{j}$$

・・・

The divergence of E equals rho, the divergence of B equals zero, the rotation of E equals minus one over c times the partial of B with respect to t, the rotation of B equals one over c times the partial of E with respect to t plus one over c times j.

・・・

電磁場 electromagnetic field は空間の各点に電場 electric field が分布したベクトル場 $\vec{E}$ と磁束密度 magnetic flux density が分布したベクトル場 $\vec{B}$ によって表されますが，これらはマックスウェル方程式と呼ばれる1階の連立偏微分方程式 partial differential equation を満たします。

・・・

第 III 部　実　践＝大学で学ぶ方程式

● 電磁場中の荷電粒子の運動方程式
equation of motion of a charged particle in the electromagnetic field

$$m\frac{d^2\vec{x}(t)}{dt^2} = e\vec{E}\left(\vec{x}(t),t\right) + \frac{e}{c}\frac{d\vec{x}(t)}{dt} \times \vec{B}\left(\vec{x}(t),t\right)$$

・・

m times d squared x of t d t squared equals e times E of x of t, t plus e over c times the vector product of d x of t d t and B of x of t, t.

・・

電場 $\vec{E}$ と磁束密度 $\vec{B}$ によって表される電磁場の影響下にある**電荷**[5] e の**荷電粒子** charged particle についてのニュートンの運動方程式に他なりません。

・・

---
[5] 電荷 electric charge

電磁気学の方程式の英語発音

●電磁ポテンシャル
electromagnetic potential

$$\vec{E} = -\vec{\nabla}\phi - \frac{1}{c}\frac{\partial \vec{A}}{\partial t}$$
$$\vec{B} = \vec{\nabla} \times \vec{A}$$

・・

E equals minus the gradient of phi minus one over c times the partial of A with respect to t, B equals the rotation of A.

・・

電磁場の電場と磁束密度を表すベクトル場は，スカラーポテンシャル scalar potential と呼ばれる電位 electric potential を表す関数 ϕ の微分とベクトルポテンシャル vector potential と呼ばれるベクトル場 $\vec{A}$ の時間微分と回転によってこのように書き表すことができます。ベクトルポテンシャルとスカラーポテンシャルを電磁ポテンシャルと呼びます。

・・

第 III 部　実　践＝大学で学ぶ方程式

●電磁ポテンシャルの波動方程式
wave equation for electromagnetic potentials

$$\Delta \vec{A} - \frac{1}{c^2}\frac{\partial^2 \vec{A}}{\partial t^2} = -\frac{1}{c}\vec{j}$$

・・・

The Laplacian of A minus one over c squared times the second partial of A with respect to t equals minus one over c times j.

・・・

電場と磁束密度がマックスウェル方程式を満たすことから，電場と磁束密度を微分や回転として表すことができる電磁ポテンシャルも何らかの方程式を満たすことになりますが，それがこの 2 階の偏微分方程式であり**波動方程式 wave equation** と呼ばれます。

・・・

量子力学の方程式の英語発音

●シュレーディンガー方程式
Schrödinger equation

$$i\hbar \frac{\partial \psi}{\partial t} = H\psi$$

....

i h bar *d* psi *d t* equals *H* psi.
....

ミクロの世界での運動を記述する方程式は発見者シュレーディンガー Schrödinger の名前をとってシュレーディンガー方程式あるいはシュレーディンガーの波動方程式と呼ばれます。

....

第 III 部　実　践＝大学で学ぶ方程式

● 1 体問題のシュレーディンガー方程式
Schrödinger equation for one-particle problem

$$i\hbar\frac{\partial}{\partial t}\psi(\vec{x},t) = -\frac{\hbar^2}{2m}\Delta\psi(\vec{x},t) + V(\vec{x},t)\psi(\vec{x},t)$$

・・・・・・・・・・・・・・・・・・・・・・・・・・・・・・・・・・・・

i h bar d d t psi of x t equals minus h bar squared over two m times the Laplacian of psi of x t plus V of x t times psi of x t.

・・・・・・・・・・・・・・・・・・・・・・・・・・・・・・・・・・・・

質量 m の粒子が位置エネルギー V による力の作用を受けて運動するとき，**量子力学** quantum mechanics ではこのようなシュレーディンガー方程式を満たす**波動関数**[6] ψ の絶対値の 2 乗がその**存在確率** existing probability を表します。

・・・・・・・・・・・・・・・・・・・・・・・・・・・・・・・・・・・・

---
[6] 波動関数 wave function

量子力学の方程式の英語発音

●正準交換関係
canonical commutation relation

$$[x_k, p_\ell] \equiv x_k p_\ell - p_\ell x_k = i\hbar \delta_{k\ell}$$

・・・・・・・・・・・・・・・・・・・・・・・・・・・・・・・・・・・・・

The commutator of x sub k and p sub ℓ is defined as x sub k times p sub ℓ minus p sub ℓ times x sub k, equals i h bar delta sub k ℓ.

・・・・・・・・・・・・・・・・・・・・・・・・・・・・・・・・・・・・・

ミクロの世界では粒子 particle の位置と運動量のかけ算はかけ算の順番を替えると違った結果となりますが、それを表すのがこの正準交換関係 canonical commutation relation です。

・・・・・・・・・・・・・・・・・・・・・・・・・・・・・・・・・・・・・

第 III 部　実　践＝大学で学ぶ方程式

● ハイゼンベルク方程式
Heisenberg equation

$$i\hbar \frac{dx_k(t)}{dt} = [x_k(t), H]$$
$$i\hbar \frac{dp_k(t)}{dt} = [p_k(t), H]$$

・・・・・・・・・・・・・・・・・・・・・・・・・・・・・・・・・・・・・

i h bar d x sub k of t d t equals the commutator of x sub k of t and H, i h bar d p sub k of t d t equals the commutator of p sub k of t and H.

・・・・・・・・・・・・・・・・・・・・・・・・・・・・・・・・・・・・・

力学の基本方程式であるハミルトンの正準方程式に対応する量子力学の基本方程式がこのハイゼンベルク方程式です。

・・・・・・・・・・・・・・・・・・・・・・・・・・・・・・・・・・・・・

量子力学の方程式の英語発音

● クラインーゴルドン方程式
Klein-Gordon equation

$$-\hbar^2 \frac{\partial^2 \psi}{\partial t^2} = -\hbar^2 c^2 \nabla^2 \psi + m^2 c^2 \psi$$

....

Minus h bar squared d squared psi $d\,t$ squared equals minus h bar squared c squared times del squared psi plus m squared c squared times psi.

....

量子力学の基本方程式であるシュレーディンガー方程式は相対性理論が考慮されていないため、エネルギーの高い**素粒子 elementary particle** の運動などを記述することはできません。そこで相対性理論を考慮して見いだされた量子力学の基本方程式がクライン-ゴルドン方程式です。これは**中間子 meson** に対する方程式として湯川秀樹博士によって取り上げられました。

....

●ディラック方程式
Dirac equation

$$i\hbar\frac{\partial \psi}{\partial t} = i\hbar c\vec{\alpha}\cdot\vec{\nabla}\psi - mc^2\beta\psi$$

・・・・・・・・・・・・・・・・・・・・・・・・・・・・・・・・・・・・・・・

i h bar d psi d t equals i h bar c times the inner product of the vector alpha and the gradient of psi minus m c squared times beta psi.

・・・・・・・・・・・・・・・・・・・・・・・・・・・・・・・・・・・・・・・

相対性理論を考慮して見いだされたクライン-ゴルドン方程式は時間についての微分が2階の波動方程式となり，シュレーディンガー方程式が時間について1階の微分しか含まない点と大きく異なっています．そのため，時間について1階の微分のみを含むような相対性理論を考慮した波動方程式が必要となりますが，それがディラック Dirac によって発見されたディラック方程式です．

・・・・・・・・・・・・・・・・・・・・・・・・・・・・・・・・・・・・・・・

場の量子論の方程式の英語発音

●場の変分原理

variational principle for fields

$$\delta \int \mathcal{L}\left(\phi(\vec{x},t), \vec{\nabla}\phi(\vec{x},t), \frac{\partial}{\partial t}\phi(\vec{x},t)\right) d^3xcdt = 0$$

・・・・・・・・・・・・・・・・・・・・・・・・・・・・・・・・・・・・

The variation of the integral of L of phi of x t, nabla phi of x t, d d t phi of x t d cubed x c d t equals zero.

・・・・・・・・・・・・・・・・・・・・・・・・・・・・・・・・・・・・

空間の各点に分布した変数としての場に対する最小作用原理が場の**変分原理** variational principle です。

・・・・・・・・・・・・・・・・・・・・・・・・・・・・・・・・・・・・

●場の変分原理（簡略記法）
variational principle for fields (brief expression)

$$\delta \int \mathcal{L}\left(\phi(x), \partial \phi(x)\right) d^4 x = 0$$

The variation of the integral of L of phi of x, the partial phi of x d to the fourth x equals zero.

場の量子論 quantum field theory では空間座標や時間座標，それに様々な場が出てくるため方程式が複雑になってしまいます。そこで，できるだけ簡略化した式の書き方が用いられます。場の変分原理もこのように書くと少し見やすくなります。

場の量子論の方程式の英語発音

●場の方程式
field equation

$$\frac{\partial \mathcal{L}}{\partial \phi} - \frac{\partial}{\partial t}\frac{\partial \mathcal{L}}{\partial \left(\frac{\partial \phi}{\partial t}\right)} - \vec{\nabla} \cdot \frac{\partial \mathcal{L}}{\partial (\vec{\nabla}\phi)} = 0$$

・・・・・・・・・・・・・・・・・・・・・・・・・・・・・・・・・・・・・

d L d phi minus d d t d L d d phi d t minus the divergence of d L d gradient phi equals zero.

・・・・・・・・・・・・・・・・・・・・・・・・・・・・・・・・・・・・・

空間に分布した変数である場の運動は最小作用原理に他ならない変分原理に従いますが，その結果として場が満たす方程式が場の方程式で，物体の運動の場合のニュートンの運動方程式に対応するものです。

・・・・・・・・・・・・・・・・・・・・・・・・・・・・・・・・・・・・・

●場の交換関係と反交換関係
commutation and anti-commutation relations for fields

$$[\phi(\vec{x},t), \phi^*(\vec{y},t)] \equiv \phi(\vec{x},t)\phi^*(\vec{y},t) - \phi^*(\vec{y},t)\phi(\vec{x},t)$$
$$= \hbar\delta^3(\vec{x}-\vec{y}),$$
$$\{\psi(\vec{x},t), \psi^*(\vec{y},t)\} \equiv \psi(\vec{x},t)\psi^*(\vec{y},t) + \psi^*(\vec{y},t)\psi(\vec{x},t)$$
$$= \hbar\delta^3(\vec{x}-\vec{y})$$

・・・

The commutator of phi of x t and phi star of y t is defined to be phi of x t times phi star of y t minus phi star of y t times phi of x t, equals h bar times delta cubed of x minus y, the anticommutator of psi of x t and psi star of y t is defined to be psi of x t times psi star of y t plus psi star of y t times psi of x t, equals h bar times delta cubed of x minus y.

・・・

場の量子論では空間の各点に分布する変数である場は，かけ算の順番を変えると答が変わってくる性質を持ちますが，これまでのところではここにある**交換関係** commutation relation か**反交換関係** anti-commutation relation を満たすものしか考えられていません。

・・・

場の量子論の方程式の英語発音

●朝永-シュウィンガー方程式
Tomonaga-Schwinger equation

$$i\hbar\frac{\delta}{\delta\sigma}|\Psi(\sigma)\rangle = H_\sigma|\Psi(\sigma)\rangle$$

・・

i h bar delta delta sigma the ket vector of upper case psi of sigma equals H sub sigma the ket vector of upper case psi of sigma.

・・

空間の各点に分布した変数である場の運動状態の時間的な変化を表す方程式で初めて相対性理論を考慮したものが朝永-シュウィンガー方程式で，朝永振一郎博士の**超多時間理論 super-many-time theory** の中で提唱されました。

・・

第 III 部 実　践＝大学で学ぶ方程式

素粒子論の方程式の英語発音

● ワインバーグ–サラム理論のラグランジアン密度
Lagrangian density of the Weinberg-Salam theory

$$\mathcal{L}_{WS} = -\frac{1}{4}F_{\mu\nu}F^{\mu\nu} - \frac{1}{4}G_{\mu\nu}G^{\mu\nu}$$
$$+\overline{\ell_e}i\gamma^\mu \left(\partial_\mu + ig\vec{W}_\mu \cdot \vec{t} + \frac{i}{2}g'yB_\mu\right)\ell_e$$

・・・・・・・・・・・・・・・・・・・・・・・・・・・・・・・・・・

L sub W S equals minus one over four times F sub mu nu F super mu nu minus one over four times G sub mu nu G super mu nu plus ℓ sub e bar i gamma super mu times d sub mu plus i g W sub mu dot t plus i over two times g prime y B sub mu times ℓ sub e.

・・・・・・・・・・・・・・・・・・・・・・・・・・・・・・・・・・

素粒子の統一理論 unified theory のうち，朝永振一郎らによって見いだされた**電磁相互作用** electromagnetic interaction の理論を原子核 nucleus から高エネルギー電子 high-energy electron やニュートリノ neutrino が放出されるベータ崩壊 beta decay に現れる**弱い相互作用** weak interaction までも含むように拡張された場の量子論における，場の変分原理に出てくる被積分関数 integrand（ラグランジアン密度 Lagrangian density と呼ばれる）です。

・・・・・・・・・・・・・・・・・・・・・・・・・・・・・・・・・・

素粒子論の方程式の英語発音

●ヒッグス粒子のラグランジアン密度
Lagrangian density for Higgs particle

$$\mathcal{L}_H = \frac{1}{2}(gv)^2 W_\mu^+ W^{-\mu} + \frac{1}{4}v^2 \left(gW_\mu^3 - g'B_\mu\right)^2$$

・・・・・・・・・・・・・・・・・・・・・・・・・・・・・・・・・・・・・・・

L sub H equals one over two times g v squared W super plus sub mu W super minus mu plus one over four v squared times g W super three sub mu minus g prime B sub mu squared.

・・・・・・・・・・・・・・・・・・・・・・・・・・・・・・・・・・・・・・・

ワインバーグ-サラム理論のラグランジアン密度を変数変換によって変形することにより導かれた場が記述する未発見の素粒子が**ヒッグス粒子 Higgs particle** ですが、ヒッグス粒子の場の方程式を変分原理から導くためのラグランジアン密度です。

・・・・・・・・・・・・・・・・・・・・・・・・・・・・・・・・・・・・・・・

第 III 部 実　践＝大学で学ぶ方程式

●小林 – 益川理論のクォーク世代間混合式
equation for the different generation quark mixture in Kobayashi and Maskawa's theory

$$\begin{pmatrix} d' \\ s' \\ b' \end{pmatrix} = \begin{pmatrix} c_1 & s_1 c_2 & s_1 s_2 \\ -s_1 c_3 & c_1 c_2 c_3 - s_2 s_3 e^{-i\delta} & c_1 s_2 c_3 + c_2 s_3 e^{-i\delta} \\ -s_1 s_3 & c_1 c_2 s_3 + s_2 c_3 e^{-i\delta} & c_1 s_2 s_3 - c_2 c_3 e^{-i\delta} \end{pmatrix} \begin{pmatrix} d \\ s \\ b \end{pmatrix}$$

・・

The three by one matrix d prime, s prime, b prime equals the three by three matrix c sub one, s sub one times c sub two, s sub one times s sub two, minus s sub one times c sub three, c sub one times c sub two times c sub three minus s sub two times s sub three times e to the minus i delta, c sub one times s sub two times c sub three plus c sub two times s sub three times e to the minus i delta, minus s sub one times s sub three, c sub one times c sub two times s sub three plus s sub two times c sub three times e to the minus i delta, c sub one times s sub two times s sub three minus c sub two times c sub three times e to the minus i delta, times the

素粒子論の方程式の英語発音

three by one matrix d, s, b.

・・・

　弱い相互作用による現象においては**時間反転** time-reversal に関する**不変性** invariance が破れていますが，**クォーク** quark の種類を増やすことによってこの現象を説明する理論が小林-益川理論です。その中で6種類（3世代）が混ざり合って相互作用することを規定するものがこの世代間混合式です。

・・・

第 III 部 実 践＝大学で学ぶ方程式

● くり込み群の方程式
equation of renormalization group

$$x\frac{\partial}{\partial x}G(x,g) - \beta(g)\frac{\partial}{\partial g}G(x,g) = 0$$

・・・・・・・・・・・・・・・・・・・・・・・・・・・・・・・・・・・・・・

x times $d\,d\,x\,G$ of $x\,g$ minus beta of g times $d\,d\,g\,G$ of $x\,g$ equals zero.

・・・・・・・・・・・・・・・・・・・・・・・・・・・・・・・・・・・・・・

場の量子論においてはパラメーター parameter としての質量定数 mass constant や相互作用定数 interaction constant についてのくり込み renormalization と呼ばれる処方を経て初めて観測される素粒子の実際の質量や電荷が与えられますが，観測されるこれらの量はパラメーターの取り方には無関係に定まるはずのものです。このことからパラメーターとしての質量定数や相互作用定数を変化させても観測される物理量を与える基礎となる場の量は不変となりますが，その事実を数式で表したものがこの方程式です。

・・・・・・・・・・・・・・・・・・・・・・・・・・・・・・・・・・・・・・

素粒子論の方程式の英語発音

●ワインバーグ–サラム理論でのワインバーグ角
Weinberg angle in the Weinberg-Salam theory

$$\sin^2 \theta_W = \frac{3}{8}\left(1 - \frac{\alpha}{4\pi}\frac{110}{9}\log\frac{M_X^2}{Q^2}\right)$$

・・・・・・・・・・・・・・・・・・・・・・・・・・・・・・・・・・・

Sine squared theta sub W equals three over eight times one minus alpha over four pi times one hundred ten over nine times log M sub X squared over Q squared.

・・・・・・・・・・・・・・・・・・・・・・・・・・・・・・・・・・・

ワインバーグ-サラム理論のラグランジアン密度を変数変換によって変形することにより導かれたラグランジアン密度の中に出てくる荷電中間ゲージ場 gauge field の電荷などの相互作用定数を全てひとつの角度パラメーター angle parameter で表すことができ，これを**ワインバーグ角** Weinberg angle と呼んでいます。

・・・・・・・・・・・・・・・・・・・・・・・・・・・・・・・・・・・

第 III 部　実　践＝大学で学ぶ方程式

流体力学の方程式の英語発音

●オイラー方程式
Euler equation

$$\frac{\partial \vec{v}}{\partial t} + \left(\vec{v} \cdot \vec{\nabla}\right)\vec{v} = -\frac{1}{\rho}\vec{\nabla}p$$

・・

$d v d t$ plus v dot nabla v equals minus one over rho times the gradient p.
・・

粘性 viscosity を完全に無視した仮想的な**流体 fluid** は**完全流体 perfect fluid** と呼ばれますが，完全流体中の各点での**速度 velocity** の分布を表す**速度場 velocity field** が満たすニュートンの運動方程式がオイラー方程式です。
・・

流体力学の方程式の英語発音

●ナヴィエー-ストークス方程式
Navier-Stokes equation

$$\frac{\partial \vec{v}}{\partial t} + \left(\vec{v} \cdot \vec{\nabla}\right) \vec{v} - \frac{\mu}{\rho} \Delta \vec{v} = -\frac{1}{\rho} \vec{\nabla} p$$

・・・・・・・・・・・・・・・・・・・・・・・・・・・・・・・・・・・・・

$d\,v\,d\,t$ plus v dot nabla v minus mu over rho times the Laplacian of v equals minus one over rho times the gradient p.

・・・・・・・・・・・・・・・・・・・・・・・・・・・・・・・・・・・・・

粘性を考慮した流体の速度場が満たすニュートンの運動方程式はナヴィエー-ストークス方程式と呼ばれています。

・・・・・・・・・・・・・・・・・・・・・・・・・・・・・・・・・・・・・

●コルトヴェーク−ド・フリース方程式
Korteweg-de Vries equation

$$\frac{\partial u}{\partial t} = \frac{3}{2}u\frac{\partial u}{\partial x} + \frac{1}{4}\frac{\partial^3 u}{\partial x^3}$$

・・・

$d\,u\,d\,t$ equals three over two times u times $d\,u\,d\,x$ plus one over four times d cubed $u\,d\,x$ cubed.

・・・

浅い運河の水面を伝わる波はソリトン soliton と呼ばれ，形がくずれないように一定の速度で伝わっていく特殊な波となります。この浅い運河のソリトン波の運動を記述する方程式がコルトヴェーク-ド・フリース方程式です。

・・・

流体力学の方程式の英語発音

●サイン–ゴルドン方程式
sine-Gordon equation

$$\frac{\partial^2 \phi(x,t)}{\partial x^2} - \frac{1}{a^2}\frac{\partial^2 \phi(x,t)}{\partial t^2} = \sin \phi(x,t)$$

....

d squared phi of x t d x squared minus one over a squared times d squared phi of x t d t squared equals sine phi of x t.

....

真空中を伝わる光は電磁場の電磁ポテンシャルについての波動方程式により記述されますが，特殊な結晶中を伝わる光の場合にはこのように**非線形項** nonlinear term として**正弦関数** sine function が加わった波動方程式により記述されます。これは形がクライン-ゴルドン方程式に似ているのでサイン-ゴルドン方程式と呼ばれています。

....

第 III 部　実　践＝大学で学ぶ方程式

●連続の方程式
equation of continuity

$$\frac{\partial \rho}{\partial t} + \vec{\nabla} \cdot (\rho \vec{v}) = 0$$

・・

d rho d t plus the divergence of rho v equals zero.

・・

　流体においては各点における流体の密度 density の変化量は速度場が表す速度で周囲からその点に流入してくる流体の量に等しくなります。この事実を数学的に表したものが連続の方程式です。

・・

古典数学の方程式の英語発音

●フェルマーの方程式
Fermat's equation

$$x^n + y^n = z^n$$

....

x to the n plus y to the n equals z to the n.
....

数年前にやっと解決した有名なフェルマー Fermat の問題に登場する方程式です。
....

第 III 部　実　践＝大学で学ぶ方程式

●コーシー‐リーマン方程式
Cauchy-Riemann equations

$$\frac{\partial u(x,y)}{\partial x} = \frac{\partial v(x,y)}{\partial y}$$
$$\frac{\partial u(x,y)}{\partial y} = -\frac{\partial v(x,y)}{\partial x}$$

・・・・・・・・・・・・・・・・・・・・・・・・・・・・・・・・・・・・・・・

d u of x y d x equals d v of x y d y, d u of x y d y equals minus d v of x y d x.

・・・・・・・・・・・・・・・・・・・・・・・・・・・・・・・・・・・・・・・

微分可能な複素関数の**実部** real part と**虚部** imaginary part が満たさなければならない方程式です。

・・・・・・・・・・・・・・・・・・・・・・・・・・・・・・・・・・・・・・・

古典数学の方程式の英語発音

●超幾何微分方程式
hypergeometric differential equation

$$z(z-1)\frac{d^2w}{dz^2} + \{(\alpha+\beta+1)z - \gamma\}\frac{dw}{dz} + \alpha\beta w = 0$$

・・・・・・・・・・・・・・・・・・・・・・・・・・・・・・・・・・・・・

z times z minus one times d squared w d z squared plus alpha plus beta plus one times z minus gamma times d w d z plus alpha times beta times w equals zero.

・・・・・・・・・・・・・・・・・・・・・・・・・・・・・・・・・・・・・

特異点 singularity を持つ 2 階同次線形常微分方程式の代表的なもので，物理学だけでなく純粋数学でも重要となる微分方程式です。

・・・・・・・・・・・・・・・・・・・・・・・・・・・・・・・・・・・・・

●積分方程式
integral equation

$$\phi(x) - \lambda \int_a^b K(x,y)\phi(y)dy = f(x)$$

....

Phi of x minus lambda times the integral from a to b of K of x y times phi of y d y equals f of x.
....

方程式の中に積分が入ってくるものは積分方程式と呼ばれます。
....

古典数学の方程式の英語発音

●コーシーの積分公式
Cauchy's integration formula

$$f(z) = \frac{1}{2\pi i} \oint_C \frac{f(\zeta)}{\zeta - z} d\zeta$$

・・・・・・・・・・・・・・・・・・・・・・・・・・・・・・・・・・・・・・

f of z equals one over two pi i times the contour integral along C of f of zeta over zeta minus z d zeta.

・・・・・・・・・・・・・・・・・・・・・・・・・・・・・・・・・・・・・・

微分可能な複素関数について成り立つ恒等式として最も有名な公式です。これによれば，微分可能な複素関数の各点での値は勝手には決めることができず，このコーシーの積分公式にあるように他の点での値によって決定されてしまいます。

・・・・・・・・・・・・・・・・・・・・・・・・・・・・・・・・・・・・・・

●作用素のスペクトル分解
spectral resolution of an operator

$$A = \int_{-\infty}^{\infty} \lambda dE_\lambda$$

....

A equals the integral from minus infinity to infinity of lambda d E sub lambda.

....

微分や積分などの操作は一般に**線形作用素** linear operator と呼ばれる数学概念だと考えられますが，それは数をかける操作に還元することができます。この数は**固有値** proper value と呼ばれます。その公式が作用素の**スペクトル分解** spectral resolution と呼ばれるものです。

....

英文索引

[a]

absolute value	96
acceleration	251
addition theorem	50, 127
analytic geometry	63
angle	110
anti-commutation relation	271
area	65
arithmetic mean	48
arithmetic series	75
associative law	23
atomic bomb	246

[b]

base	61, 65
beta decay	273
bijection	37
binomial theorem	124
brace	24
bracket	24

[c]

canonical commutation relation	264
cap	23
Cartesian product	32
Cauchy-Riemann equations	285
Cauchy's integration formula	288
Cayley–Hamilton's theorem	119
center	68
charged particle	259
circle	68
column	110
combination	121
common logarithm	61
commutation relation	271
commutative law	23
complement	26
complementary event	126
complex function	39
complex number	31
component	110
composite function	86
composite mapping	38
conjunction	41
conservation law	252
constant	95
continuous	101
coordinates	63
correspondence	35
cubic equation	48
cup	23
curvature tensor	249

[d]

De Morgan's formula	27, 44
definite integral	95
density	283
determinant	118
diameter	68
difference	25
differential	83
differential equation	251

differential operation	85	equivalent	40
differentiation	83	Euler equation	279
Dirac	267	Euler-Lagrange equation	253
Dirac equation	267	Euler's formula	57
disjunction	41	event	126
distance	66	exclusive event	127
distributive law	25	existing probability	263
divergence	257	expectation	130
domain	69	exponential	55
		exponential function	54

[e]

edge	63		
Einstein	246		

[f]

Einstein's equation of gravity field		factorial	77
	249	factorization	48
Einstein's formula	246	Fermat	284
electric charge	259	Fermat's equation	284
electric field	258	Fibonacci series	81
electric potential	260	field	255
electromagnetic field	258	field equation	270
electromagnetic interaction	273	figure	68
electromagnetic potential	260	finite set	28
element	16	fluid	279
elementary function	46	force	251
elementary particle	266	fourth order equation	48
ellipse	69	free fall	248
empty set	17	function	35, 68
energy	246	fundamental theorem of calculus	
energy conservation law	252		103
energy-momentum tensor	249		

[g]

equal	18		
equation	67	gauge field	278
equation of continuity	283	general term	75
equation of motion	251	general theory of relativity	247
		geodesics	248

291

geodesics equation	248
geometric mean	48
geometric series	76
geometry	63
gradient	251
graph	70
gravitational constant	249
gravity	247
gravity field	247
Greek letters	28

[h]

Hamilton-Jacobi equation	255
Hamilton's canonical equation	254
Hamilton's principle of least action	252
height	65
Heisenberg equation	265
Heron's formula	66
Higgs particle	274
high-energy electron	273
hyperbola	69
hyperbolic function	57
hypergeometric differential equation	286

[i]

identity	47
image	35
imaginary part	285
imaginary unit	57
implication	41
indefinite integral	95
independent event	129
inequality	48
inertial system	245
injection	37
inner product	109
integer	30
integral	95
integral equation	287
integration	95
integration by parts	104
interaction constant	277
interior angle	63
intersection	22
interval	101
invariance	276
inverse mapping	37
inverse matrix	118
inverse operation	95
irrational number	31

[k]

kinetic energy	252
Klein-Gordon equation	266
Korteweg-de Vries equation	281

[l]

Lagrangian	252
Lagrangian density	273
Laplacian	91
law	89
law of cosine	64
law of exponent	54
law of sines	64
left-hand side	18
Leibnizian style	83

length	63
limit	83
linear equation	72
linear operation	85
linear operator	289
linear transformation	111
linearity	102
logarithmic function	54, 61
logical operation	40
logical product	41
logical sum	40
Lorentz transformation	245
lower case	18

[m]

magnetic flux density	258
mapping	35
mass	246, 251
mass constant	277
mathematics	16
matrix	108
matter	246
Maxwell equation	258
mean value	130
meson	266
metric tensor	247
momentum	249
momentum conservation law	252
multiplication	109
multiplication theorem	129
multivariable	89
multivariable function	89

[n]

nabla	93
natural logarithm	61
natural number	30
Navier-Stokes equation	280
negation	41
neutrino	273
Newton's equation of motion	251
nonlinear term	282
normal probability density function	132
n-th order differencial	85
nucleus	273
n-variable function	89

[o]

one-particle problem	263
origin	70
orthogonal	70
outer product	114

[p]

parabola	69
parallelogram	115
parameter	277
parenthesis	24
partial differential	89
partial sum	75
path	252
perfect fluid	279
permutation	121
perpendicular	68
physics	251

plane	72	right-hand side	18
plane geometry	71	right triangle	64
point	66	rotation	119, 257
polynomial function	49	round bracket	24
polynomial theorem	124	row	110
position vector	109		
potential energy	251		

[s]

power set	33	scalar curvature	249
primitive function	95	scalar multiplication	109
principle	89	scalar potential	260
probability	121	scalar product	109
probability density function	132	Schrödinger	262
proper value	289	Schrödinger equation	262
proposition	40	Schwarz' inequality	104
Pythagorean theorem	64	Schwarzschild	250
		Schwarzschild's metric	250
		second order differential	84

[q]

quadratic equation	46	sequence	74
quadratic function	46	series	74
quantum field theory	269	set	16
quantum mechanics	263	simultaneous equations	71
quark	276	sine function	282
		sine-Gordon equation	282
		singularity	286

[r]

		soliton	281
radian	110	space	34, 257
radius	68	space-time coordinates	245, 247
random variable	130	space-time metric	247
rational number	31	space-time metric form	247
real function	39	special theory of relativity	246
real number	31	spectral resolution	289
real part	285	sphere	72
recurrence formula	75	square bracket	24
renormalization	277	square matrix	116
right angle	64		

standard deviation	131	union	21
statistics	121	upper case	17
straight line	67		

[v]

sub set	20		
suffix	22	variable	89
super	27	variance	131
super-many-time theory	272	variation	252
superscript	27	variational principle	268
surjection	36	variational principle for fields	268
symmetric difference	26	vector	108
system of rectangular coordinates	66	vector field	257
		vector potential	260
system of space-time coordinates	245	vector product	114
		velocity	279
system of spacial coordinates	70	velocity field	279
		vertex	63
		viscosity	279

[t]

[w]

term	74		
theory of relativity	247	wave equation	261
third order differential	84	wave function	263
time-reversal	276	weak interaction	273
Tomonaga-Schwinger equation	272	Weinberg angle	278
triangle	63	Weinberg-Salam theory	273
trigonometric function	50		

[u]

[x]

unified theory	273	x axis	66
uniform motion	245		

和文索引

【英数字】

1 次方程式 linear equation	72
1 体問題 one-particle problem	263
2 階微分 second order differential	84
2 項定理 binomial theorem	124
2 次関数 quadratic function	46
2 次方程式 quadratic equation	46
3 階微分 third order differential	84
3 次方程式 cubic equation	48
4 次方程式 fourth order equation	48
x 軸 x axis	66
n 階微分 n-th order differential	85
n 変数関数 n-variable function	89

【あ行】

アインシュタイン Einstein	246
アインシュタインの公式 Einstein's formula	246
アインシュタインの重力場方程式 Einstein's equation of gravity field	249
位置エネルギー potential energy	251
位置ベクトル position vector	109
一般項 general term	75
一般相対性理論 general theory of relativity	247
因数分解 factorization	48
右辺 right-hand side	18
運動エネルギー kinetic energy	252
運動経路 path	252
運動方程式 equation of motion	251
運動量 momentum	249
運動量保存則 momentum conservation law	252
エネルギー energy	246
エネルギー・運動量テンソル energy-momentum tensor	249
エネルギー保存則 energy conservation law	252
円 circle	68
オイラーの公式 Euler's formula	57
オイラー方程式 Euler equation	279
オイラー−ラグランジュ方程式 Euler-Lagrange equation	253
大文字 upper case	17

【か行】

階乗 factorial	77
外積 outer product	114
解析学の基本定理 the fundamental theorem of calculus	103
解析幾何 analytic geometry	63
回転 rotation	119, 257
角度 angle	110
確率 probability	121
確率変数 random variable	130
確率密度関数 probability density function	132
かけ算 multiplication	109
加速度 acceleration	251

和文索引

括弧 parenthesis	24
荷電粒子 charged particle	259
加法定理 addition theorem	50, 127
含意 implication	41
関数 function	35, 68
慣性系 inertial system	245
完全流体 perfect fluid	279
幾何学 geometry	63
期待値 expectation	130
逆演算 inverse operation	95
逆行列 inverse matrix	118
逆写像 inverse mapping	37
球 sphere	72
級数 series	74
行 row	110
行列 matrix	108
行列式 determinant	118
極限 limit	83
曲率テンソル curvature tensor	249
虚数単位 imaginary unit	57
虚部 imaginary part	285
距離 distance	66
ギリシャ文字 Greek letters	28
空間 space	34, 257
空間座標系 system of spacial coordinates	70
空集合 empty set	17
クォーク quark	276
区間 interval	101
組み合わせ combination	121
クライン－ゴルドン方程式 Klein-Gordon equation	266
グラフ graph	70
くり込み renormalization	277
計量テンソル metric tensor	247
ゲージ場 gauge field	278
ケーリー－ハミルトンの定理 Cayley–Hamilton's theorem	119
結合法則 associative law	23
原子核 nucleus	273
原始関数 primitive function	95
原点 origin	70
原爆 atomic bomb	246
原理 principle	89
項 term	74
高エネルギー電子 high-energy electron	273
交換関係 commutation relation	271
交換法則 commutative law	23
合成関数 composite function	86
合成写像 composite mapping	38
恒等式 identity	47
勾配 gradient	251
コーシーの積分公式 Cauchy's integration formula	288
コーシー－リーマン方程式 Cauchy-Riemann equations	285
小文字 lower case	18
固有値 proper value	289
コルトヴェーク－ド・フリース方程式 Korteweg-de Vries equation	281

【さ行】

差 difference	25
サイン－ゴルドン方程式 sine-Gordon equation	282
座標 coordinates	63

和文索引

左辺 left-hand side	18
三角関数 trigonometric function	50
三角形 triangle	63
時間反転 time-reversal	276
時空計量 space-time metric	247
時空計量の式 space-time metric form	247
時空座標 space-time coordinates	245, 247
時空座標系 system of space-time coordinates	245
事象 event	126
指数関数 exponential function	54
指数法則 law of exponent	54
自然数 natural number	30
自然対数 natural logarithm	61
磁束密度 magnetic flux density	258
実関数 real function	39
実数 real number	31
実部 real part	285
質量 mass	246, 251
質量定数 mass constant	277
写像 mapping	35
集合 set	16
集合積 intersection	22
集合和 union	21
自由落下 free fall	248
重力 gravity	247
重力定数 gravitational constant	249
重力場 gravity field	247
シュレーディンガー Schrödinger	262
シュレーディンガー方程式 Schrödinger equation	262
シュワルツシルト Schwarzschild	250
シュワルツシルト計量 Schwarzschild's metric	250
シュワルツの不等式 Schwarz' inequality	104
順列 permutation	121
乗法定理 multiplication theorem	129
常用対数 common logarithm	61
初等関数 elementary function	46
垂線 perpendicular	68
数学 mathematics	16
数列 sequence	74
スカラー曲率 scalar curvature	249
スカラー積 scalar product	109
スカラー倍 scalar multiplication	109
スカラーポテンシャル scalar potential	260
図形 figure	68
スペクトル分解 spectral resolution	289
正規確率密度関数 normal probability density function	132
正弦関数 sine function	282
正弦定理 law of sines	64
正準交換関係 canonical commutation relation	264
整数 integer	30
成分 component	110
正方行列 square matrix	116
積分 integral	95
積分法 integration	95

積分方程式 integral equation	287
絶対値 absolute value	96
漸化式 recurrence formula	75
線形演算 linear operation	85
線形作用素 linear operator	289
線形性 linearity	102
線形変換 linear transformation	111
全射 surjection	36
全単射 bijection	37
像 image	35
相加平均 arithmetic mean	48
双曲線 hyperbola	69
双曲線関数 hyperbolic function	57
相互作用定数 interaction constant	277
相乗平均 geometric mean	48
相対性理論 theory of relativity	247
添字 suffix	22
測地線 geodesics	248
測地線の方程式 geodesics equation	248
速度 velocity	279
速度場 velocity field	279
ソリトン soliton	281
素粒子 elementary particle	266
存在確率 existing probability	263

【た行】

対応 correspondence	35
対称差 symmetric difference	26
対数関数 logarithmic function	54, 61
楕円 ellipse	69
高さ height	65
多項式関数 polynomial function	49
多項定理 polynomial theorem	124
多変数 multivariable	89
多変数関数 multivariable function	89
単射 injection	37
力 force	251
中間子 meson	266
中心 center	68
超幾何微分方程式 hypergeometric differential equation	286
超多時間理論 super-many-time theory	272
頂点 vertex	63
直積 direct product	32
直線 straight line	67
直角 right angle	64
直角三角形 right triangle	64
直径 diameter	68
直交 orthogonal	70
直交座標系 system of rectangular coordinates	66
底 base	61
定数 constant	95
定積分 definite integral	95
底辺 base	65
ディラック Dirac	267
ディラック方程式 Dirac equation	267
点 point	66
電位 electric potential	260
電荷 electric charge	259
電磁相互作用 electromagnetic interaction	273

和文索引

電磁場 electromagnetic field 258
電磁ポテンシャル
   electromagnetic potential 260
電場 electric field 258
統一理論 unified theory 273
統計 statistics 121
等号 equal 18
等差級数 arithmetic series 75
等速直線運動 uniform motion 245
同値 equivalent 40
等比級数 geometric series 76
特異点 singularity 286
特殊相対性理論
   special theory of relativity 246
独立事象 independent event 129
朝永-シュウィンガー方程式
   Tomonaga-Schwinger equation 272
ド・モルガンの公式
   De Morgan's formula 27, 44

【な行】

内角 interior angle 63
内積 inner product 109
ナヴィエ-ストークス方程式
   Navier-Stokes equation 280
長さ length 63
ナブラ nabla 93
ニュートリノ neutrino 273
ニュートンの運動方程式 Newton's
   equation of motion 251
粘性 viscosity 279

【は行】

場 field 255
ハイゼンベルク方程式
   Heisennberg equation 265
排反事象 exclusive event 127
発散 divergence 257
波動関数 wave function 263
波動方程式 wave equation 261
場の変分原理
   variational principle for fields 268
場の方程式 field equation 270
場の量子論 quantum field theory 269
ハミルトンの最小作用原理
   Hamilton's principle of least action 252
ハミルトンの正準方程式
   Hamilton's canonical equation 254
ハミルトン-ヤコビ方程式
   Hamilton-Jacobi equation 255
パラメーター parameter 277
半径 radius 68
反交換関係
   anti-commutation relation 271
非線形項 nonlinear term 282
ピタゴラスの定理
   Pythagorean theorem 64
ヒッグス粒子 Higgs particle 274
否定 negation 41
微分 differential 83
微分操作 differential operation 85
微分法 differentiation 83

微分方程式 differential equation	251	偏微分 partial differential	89
標準偏差 standard deviation	131	変分 variation	252
フィボナッチ数列 Fibonacci series	81	変分原理 variational principle	268
フェルマー Fermat	284	法則 law	89
フェルマーの方程式 Fermat's equation	284	方程式 equation	67
		放物線 parabola	69
複素関数 complex function	39	補集合 complement	26
複素数 complex number	31	保存法則 conservation law	252

【ま行】

物体 matter	246
物理 physics	251
不定積分 indefinite integral	95
不等式 inequality	48
部分集合 subset	20
部分積分 integration by parts	104
部分和 partial sum	75
不変性 invariance	276
分散 variance	131
分配法則 distributive law	25
平均値 mean value	130
平行四辺形 parallelogram	115
平面 plane	72
平面幾何 plane geometry	71
ベータ崩壊 beta decay	273
べき集合 power set	33
ベクトル vector	108
ベクトル積 vector product	114
ベクトル場 vector field	257
ベクトルポテンシャル vector potential	260
ヘロンの公式 Heron's formula	66
辺 edge	63
変数 variable	89

マックスウェル方程式 Maxwell equation	258
密度 density	283
無理数 irrational number	31
命題 proposition	40
面積 area	65

【や行】

有限集合 finite set	28
有理数 rational number	31
要素 element	16
余弦定理 law of cosine	64
余事象 complementary event	126
弱い相互作用 weak interaction	273

【ら行】

ライプニッツ流 Leibnizian style	83
ラグランジアン Lagrangian	252
ラグランジアン密度 Lagrangian density	273
ラジアン radian	110
ラプラシアン Laplacian	91
流体 fluid	279
領域 domain	69

和文索引

量子力学 quantum mechanics	263
列 column	110
連続 continuous	101
連続の方程式 equation of continuity	283
連立方程式 simultaneous equations	71
ローレンツ変換 Lorentz transformation	245
論理演算 logical operation	40
論理積 logical product	41
論理和 logical sum	40

【わ行】

ワインバーグ角 Weinberg angle	278
ワインバーグ-サラム理論 Weinberg-Salam theory	273

N.D.C.410.7　　302p　　18cm

ブルーバックス　B-1366

数学版　これを英語で言えますか？
Let's speak mathematics!

2002年 4月20日　　第 1 刷発行
2024年12月13日　　第14刷発行

著者	保江邦夫
監修者	エドワード・ネルソン
発行者	篠木和久
発行所	株式会社講談社
	〒112-8001 東京都文京区音羽2-12-21
電話	出版　03-5395-3524
	販売　03-5395-5817
	業務　03-5395-3615
印刷所	(本文表紙印刷) 株式会社ＫＰＳプロダクツ
	(カバー印刷) 信毎書籍印刷株式会社
製本所	株式会社ＫＰＳプロダクツ

定価はカバーに表示してあります。
©保江邦夫　2002, Printed in Japan
落丁本・乱丁本は購入書店名を明記のうえ、小社業務宛にお送りください。送料小社負担にてお取替えします。なお、この本についてのお問い合わせは、ブルーバックス宛にお願いいたします。
本書のコピー、スキャン、デジタル化等の無断複製は著作権法上での例外を除き禁じられています。本書を代行業者等の第三者に依頼してスキャンやデジタル化することはたとえ個人や家庭内の利用でも著作権法違反です。
Ⓡ〈日本複製権センター委託出版物〉複写を希望される場合は、日本複製権センター（電話03-6809-1281）にご連絡ください。

ISBN4-06-257366-0

発刊のことば

科学をあなたのポケットに

二十世紀最大の特色は、それが科学時代であるということです。科学は日に日に進歩を続け、止まるところを知りません。ひと昔前の夢物語もどんどん現実化しており、今やわれわれの生活のすべてが、科学によってゆり動かされているといっても過言ではないでしょう。

そのような背景を考えれば、学者や学生はもちろん、産業人も、セールスマンも、ジャーナリストも、家庭の主婦も、みんなが科学を知らなければ、時代の流れに逆らうことになるでしょう。

ブルーバックス発刊の意義と必然性はそこにあります。このシリーズは、読む人に科学的に物を考える習慣と、科学的に物を見る目を養っていただくことを最大の目標にしています。そのためには、単に原理や法則の解説に終始するのではなくて、政治や経済など、社会科学や人文科学にも関連させて、広い視野から問題を追究していきます。科学はむずかしいという先入観を改める表現と構成、それも類書にないブルーバックスの特色であると信じます。

一九六三年九月

野間省一